Infrastructure as Code for Beginners

Deploy and manage your cloud-based services with Terraform and Ansible

Russ McKendrick

BIRMINGHAM—MUMBAI

Infrastructure as Code for Beginners

Group Product Manager: Preet Ahuja

Publishing Product Manager: Surbhi Suman

Senior Content Development Editor: Adrija Mitra

Technical Editor: Nithik Cheruvakodan

Copy Editor: Safis Editing

Project Coordinator: Ashwin Kharwa

Proofreader: Safis Editing

Indexer: Subalakshmi Govindhan

Production Designer: Prashant Ghare

Marketing Coordinator: Agnes D'souza

First published: May 2023

Production reference: 1110523

Published by Packt Publishing Ltd.

Livery Place

35 Livery Street

Birmingham

B3 2PB, UK.

978-1-83763-163-6

www.packtpub.com

Contributors

About the author

Russ McKendrick is an experienced DevOps practitioner and system administrator with a passion for automation and containers. He has been working in IT and related industries for the better part of 30 years. During his career, he has had responsibilities in many different sectors, including first-line, second-line, and senior support in client-facing and internal teams for small and large organizations.

He works almost exclusively with Linux, using open source systems and tools across dedicated hardware and virtual machines hosted in public and private clouds at Node4, where he holds the title of practice manager (SRE and DevOps). He also buys way too many records!

I would like to thank my family and friends for their support and for being so understanding about all of the time I have spent in front of the computer writing. I would also like to thank my colleagues at Node4 and our customers for their kind words of support and encouragement throughout the writing process.

About the reviewer

Adam Hooper has nearly a decade of experience in cloud computing and has held various roles within MSPs and CSPs. Adam started as a support technician initially supporting Citrix and Hyper-V environments and then moved to provide support to customers' critical Azure environments. He now works as a platform engineer for Node4, where he spends most of his time deploying resources with Terraform and Azure DevOps.

Outside of work, Adam enjoys motorcycles, spending time with his wife, Hannah, and their two cats, spending time with family, and drinking copious amounts of whiskey.

I would like to thank Russ McKendrick for giving me the opportunity to review this book. It's been a great experience and hopefully the first of many.

Table of Contents

Preface   ix

Part 1: The Foundations – An Introduction to Infrastructure as Code

1

Choosing the Right Approach – Declarative or Imperative   3

The challenges of managing
infrastructure manually   3
My own journey   4
Today's challenges   5
Conclusion   7

What is meant by declarative and
imperative?   8
Basic Infrastructure-as-Code project   8
Declarative approach   9

Imperative approach   11
Pets versus cattle   13
Pets   13
Cattle   13
Conclusion   14

What does all this mean for our
Infrastructure-as-Code deployments?   15
Summary   16
Further reading   17

2

Ansible and Terraform beyond the Documentation   19

What is important when choosing
a tool?   19
Deployment types   20
Infrastructure and configuration   21
External interactions and secrets   21

Ease of use   22
Summary   22

Introducing Terraform   22
An HCL example – creating a resource group 23
Adding more resources   25

Introducing Ansible   29   Introducing Visual Studio Code   34
An Ansible example   30   Summary   35
                                               Further reading   36

3

Planning the Deployment   37

Planning the deployment of   Performing deployment tasks   40
our workload   37   Introducing cloud-init   42
How to approach the deployment of   Exploring the high-level architecture 43
our infrastructure   38   Summary   45
Deployment considerations   39   Further reading   45

Part 2: Getting Hands-On with the Deployment

4

Deploying to Microsoft Azure   49

Technical requirement   49   Creating a resource group   54
Introducing and preparing our   Networking   56
cloud environment   50   Ansible – reviewing the code and
Preparing our cloud environment   deploying our infrastructure   73
for deployment   50   Ansible Playbook roles overview   74
Producing the low-level design   51   Running the Ansible Playbook   81
Terraform – writing the code and   Summary   82
deploying our infrastructure   52   Further reading   82
Setting up the Terraform environment 53

5

Deploying to Amazon Web Services   85

Technical requirements   86   Preparing our cloud environment for
Introducing Amazon Web Services   86   deployment   87

Producing the low-level design   88
Ansible – writing the code and
deploying our infrastructure   89
Ansible playbook roles   91
Running the Ansible playbook   102

Terraform – reviewing the code and
deploying our infrastructure   107
Walk-through of Terraform files   107
Deploying the environment   113

Summary   114
Further reading   115

6

Building upon the Foundations   117

Understanding cloud-agnostic tools   117
Understand the differences between
our Microsoft Azure and Amazon
Web Services deployments   118
General   119
Network   119
Storage   120
Virtual machine (admin)   121
Virtual machines with scaling (web)   122
Seeing it in action   122

Understanding the differences
between our Terraform and Ansible
deployments   126
Introducing more variables   129
Making the code more reusable   131
Pop quiz   134
Summary   134
Further reading   135
Answers   135

Part 3: CI/CD and Best Practices

7

Leveraging CI/CD in the Cloud   139

Technical requirements   140
Introducing GitHub Actions   140
Running Terraform using
GitHub Actions   141
Terraform state files   141
GitHub Actions   142

Running Ansible using
GitHub Actions   155
Security best practices   158
Pop quiz   159
Summary   159
Further reading   160
Answers   160

8

Common Troubleshooting Tips and Best Practices   161

Technical requirements	161
Infrastructure as Code – best practices and troubleshooting	162
General IaC best practices	162
General IaC troubleshooting tips	164
Terraform – best practices and troubleshooting	165
Terraform – best practices	165

Terraform – troubleshooting	167
Ansible – best practices and troubleshooting	168
Ansible – best practices	168
Ansible – troubleshooting	172
Summary	173

9

Exploring Alternative Infrastructure-as-Code Tools   175

Technical requirements	175
Getting hands-on with Pulumi	176
Using Pulumi and YAML	176
Using Pulumi and Python	181
Getting hands-on knowledge of Azure Bicep	185
Working through the Bicep file	185
Deploying the Bicep file	187

Getting hands-on with AWS CloudFormation	188
AWS CloudFormation template	189
Using the AWS CLI to deploy	190
Using the AWS Management Console to deploy	190
Summary	194
Further reading	196

Index   197

Other Books You May Enjoy   204

Preface

Welcome to *Infrastructure as Code for Beginners*, your guide to managing and deploying your infrastructure through code. This book will equip you with a strong foundation in the essential concepts, tools, and techniques necessary to succeed in this ever-changing landscape.

Throughout the chapters, you will gain hands-on experience with popular Infrastructure-as-Code tools such as Terraform and Ansible, learn how to plan and deploy resources across leading public cloud providers such as Microsoft Azure and Amazon Web Services, and delve into best practices and troubleshooting strategies to help you overcome challenges and optimize your deployments.

As you progress through the book, you'll also explore the role of Continuous Integration and **Continuous Deployment (CI/CD)** in automating your Infrastructure-as-Code projects using GitHub Actions. You'll learn how to leverage CI/CD to create consistent and reliable deployments and implement security practices to secure your deployments.

Furthermore, you'll expand your Infrastructure-as-Code toolset by exploring alternative tools such as Pulumi, Azure Bicep, and AWS CloudFormation, enhancing your understanding of provider-specific and cloud-agnostic options.

By the end of this journey, you'll have gained the knowledge and confidence to plan, build, deploy, and manage your Infrastructure-as-Code projects enabling you to create efficient, scalable, and reliable infrastructure solutions that will support your projects and career for years to come.

Who this book is for

This book is tailored for Developers and System Administrators who have experience of manually deploying resources to host their applications but now want to increase their skills by automating the creation and management of infrastructure alongside their applications.

This book will empower readers who, while having a great deal of experience deploying and configuring their resources manually, now want to streamline their processes, enhance efficiency and consistency, and integrate Infrastructure as Code into their daily workflows.

What this book covers

In *Chapter 1, Choosing the Right Approach – Declarative or Imperative*, we introduce fundamental Infrastructure-as-Code concepts, covering the challenges of manual infrastructure management, declarative versus imperative approaches, and the *pets versus cattle* analogy, setting the stage for understanding its importance in modern deployments.

In *Chapter 2, Ansible and Terraform beyond the Documentation*, we delve into Terraform, an Infrastructure-as-Code tool by HashiCorp, and Ansible, a configuration management tool by Red Hat. We will explore tool selection criteria, be introduced to Terraform and Ansible, and review some guidance on using Visual Studio Code as an IDE for writing code, including some recommended extensions.

In *Chapter 3, Planning the Deployment*, we highlight the importance of planning in your Infrastructure-as-Code deployments. You will be introduced to the workload to be deployed, learn about deployment approaches, and explore a step-by-step guide for efficient and error-free execution. We will also examine the high-level infrastructure architecture, preparing us for our upcoming Azure and AWS deployment.

In *Chapter 4, Deploying to Microsoft Azure, we* explore deploying our project to Microsoft Azure, one of the two major public cloud providers covered in this book. Topics include introducing Azure, preparing the cloud environment, creating the low-level design, and using Terraform and Ansible for writing and deploying the infrastructure code.

In *Chapter 5, Deploying to Amazon Web Services, we* move onto deploying the project to **Amazon Web Services (AWS)** while highlighting the key differences between Azure and AWS. We will delve deeper into Ansible for deployment and gain insights into using both Ansible and Terraform for managing AWS resources. By the end of the chapter, you will understand how to adapt your deployment approach for different cloud providers.

In *Chapter 6, Building upon the Foundations, we delve* into the nuances of deploying high-level designs across public cloud providers using cloud-agnostic tools such as Terraform and Ansible. We will learn from my experiences of addressing variations between providers, exploring practical approaches for creating repeatable deployment processes, and looking at the importance of modular code, allowing for streamlined deployment efforts and code reusability.

In Chapter 7, Leveraging CI/CD in the Cloud, we will now focus on using CI/CD to automate infrastructure deployment. We will explore GitHub Actions, a popular CI/CD tool, and learn how to use it to run Terraform and Ansible code for both Azure and AWS.

In Chapter 8, Common Troubleshooting Tips and Best Practices, we will learn essential strategies for planning, writing, and troubleshooting Infrastructure-as-Code projects. The chapter covers best practices and troubleshooting tips for Infrastructure as Code in general and specific guidance for Terraform and Ansible. By understanding the unique challenges associated with each tool, you will be better prepared to handle obstacles that may arise during your Infrastructure-as-Code journey.

In Chapter 9, Exploring Alternative Infrastructure-as-Code Tools, we will expand our Infrastructure-as-Code toolset by exploring three additional tools: Pulumi, Azure Bicep, and AWS CloudFormation. The chapter aims to provide hands-on understanding and knowledge of these tools, highlighting the differences between cloud-agnostic and provider-specific tools and showcasing Pulumi's unique approach using familiar programming languages.

To get the most out of this book

Before starting, you should understand how you would approach and deploy the infrastructure to support one or more of your existing applications. Ideally, this infrastructure would be hosted in either Microsoft Azure or Amazon Web Services.

While this is optional, the book has been written with the assumption that you have a basic understanding of either Microsoft Azure or Amazon Web Services and some of the core services' basic concepts, such as networking or virtual machines. While this is not essential, you may have to do some further reading about why we approached tasks in a certain way regarding some of the more hands-on chapters.

We also assume that you would like to follow along with some of the more practical chapters; therefore, having credits or access to a Microsoft Azure or Amazon Web Services account that isn't used to host production resources is a bonus.

Software/hardware covered in the book	Operating system requirements
Terraform	Windows, macOS, or Linux
Ansible	Windows, macOS, or Linux
The Microsoft Azure CLI and portal	Windows, macOS, or Linux
The Amazon Web Services CLI and portal	Windows, macOS, or Linux
Pulumi	Windows, macOS, or Linux
Visual Studio Code	Windows, macOS, or Linux

There are links to the installation instructions for each tool in the further reading section of the respective chapters.

If you are using the digital version of this book, we advise you to type the code yourself or access the code from the book's GitHub repository (a link is available in the next section). Doing so will help you avoid any potential errors related to the copying and pasting of code.

Finally, feel free to experiment; check out the code accompanying this book and make changes. However, keep an eye on spending and always ensure that you terminate any resources you deploy after you have finished to avoid any unexpected and unwanted costs.

Download the example code files

You can download the example code files for this book from GitHub at `https://github.com/PacktPublishing/Infrastructure-as-Code-for-Beginners`. If there's an update to the code, it will be updated in the GitHub repository.

We also have other code bundles from our rich catalog of books and videos available at `https://github.com/PacktPublishing/`. Check them out!

Download the color images

We also provide a PDF file that has color images of the screenshots and diagrams used in this book. You can download it here: `https://packt.link/uvP61`.

Conventions used

There are a number of text conventions used throughout this book.

`Code in text`: Indicates code words in text, database table names, folder names, filenames, file extensions, pathnames, dummy URLs, user input, and Twitter handles. Here is an example: "The first three lines, `name`, `runtime`, and `description`, all define some basic meta information about our deployment."

A block of code is set as follows:

```
name: pulumi-yaml
runtime: yaml
description: A minimal Azure Native Pulumi YAML program
outputs:
  primaryStorageKey: ${storageAccountKeys.keys[0].value}
```

Any command-line input or output is written as follows:

```
$ pulumi up -c Pulumi.dev.yaml
```

Bold: Indicates a new term, an important word, or words that you see onscreen. For instance, words in menus or dialog boxes appear in **bold**. Here is an example: "The final step is to review your stack before clicking the **Submit** button, triggering the stack creation."

> **Tips or important notes**
> Appear like this.

Get in touch

Feedback from our readers is always welcome.

General feedback: If you have questions about any aspect of this book, email us at customercare@packtpub.com and mention the book title in the subject of your message.

Errata: Although we have taken every care to ensure the accuracy of our content, mistakes do happen. If you have found a mistake in this book, we would be grateful if you would report this to us. Please visit www.packtpub.com/support/errata and fill in the form.

Piracy: If you come across any illegal copies of our works in any form on the internet, we would be grateful if you would provide us with the location address or website name. Please contact us at copyright@packt.com with a link to the material.

If you are interested in becoming an author: If there is a topic that you have expertise in and you are interested in either writing or contributing to a book, please visit authors.packtpub.com.

Share Your Thoughts

Once you've read *Infrastructure as Code for Beginners*, we'd love to hear your thoughts! Scan the QR code below to go straight to the Amazon review page for this book and share your feedback.

https://packt.link/r/1837631638

Your review is important to us and the tech community and will help us make sure we're delivering excellent quality content.

Download a free PDF copy of this book

Thanks for purchasing this book!

Do you like to read on the go but are unable to carry your print books everywhere? Is your eBook purchase not compatible with the device of your choice?

Don't worry, now with every Packt book you get a DRM-free PDF version of that book at no cost.

Read anywhere, any place, on any device. Search, copy, and paste code from your favorite technical books directly into your application.

The perks don't stop there, you can get exclusive access to discounts, newsletters, and great free content in your inbox daily

Follow these simple steps to get the benefits:

1. Scan the QR code or visit the link below

https://packt.link/free-ebook/978-1-83763-163-6

2. Submit your proof of purchase
3. That's it! We'll send your free PDF and other benefits to your email directly

Part 1:
The Foundations –
An Introduction to
Infrastructure as Code

In this part, we will discuss how you can approach your Infrastructure-as-Code journey, using my own as a starting point, before exploring some of the core principles of infrastructure as code.

We will also look at the tools we will be using to deploy our example workload and planning the deployment of the workload itself.

This part has the following chapters:

- *Chapter 1, Choosing the Right Approach – Declarative or Imperative*
- *Chapter 2, Ansible and Terraform beyond the Documentation*
- *Chapter 3, Planning the Deployment*

1

Choosing the Right Approach – Declarative or Imperative

Welcome to the first chapter of *Infrastructure as Code for Beginners*. In this book, we will be going on a journey that will take you through your first Infrastructure-as-Code deployment, and it is an honor to be accompanying you.

Before we dive into the tools we will be using throughout the book, we are first going to discuss some of the key concepts to try and get an understanding of the problems that you could try to solve by introducing Infrastructure as Code into your deployments.

We will be covering the following topics:

- The challenges of managing infrastructure manually
- What is meant by declarative and imperative?
- Pets versus cattle
- What does all this mean for our Infrastructure-as-Code deployments?

The challenges of managing infrastructure manually

Before we look at some of the challenges you may be facing, I quickly wanted to take you through my journey with Infrastructure as Code before it was really what we now know as Infrastructure as Code.

When I talk about Infrastructure as Code, I mean the following:

Infrastructure as Code is an approach to infrastructure management where it is provisioned and managed using code and automation tools rather than manually configuring resources through a user interface.

This allows you to version control, track, and manage your infrastructure in the same way you do with application code and, in many cases, use the same tooling, processes, and procedures you already have in place.

Infrastructure as Code can help improve your infrastructure operations' efficiency, reliability, and reproducibility by introducing consistency across your deployments and reducing deployment times versus more traditional manual deployments.

My own journey

I have been working with servers of all types for longer than I care to remember; back when I first started working with servers, it was all very much a manual process to do pretty much anything.

The bare-metal days

Before *virtualization* became a common practice, I remember having to block out a whole day to build a customer's server. This process would generally start with ensuring that the hardware I was given to work with was of the correct specification – if for some reason it wasn't, which was quite common, then I would typically have to replace RAM and hard drives, and so on.

This was to ensure that I didn't get too far into configuring the server, only to find that I had to tear it down and start from scratch; once the hardware was confirmed as being correct, it was time to start on the build itself.

Typically, to build the server, I sat in a tiny, hot, and noisy build room surrounded by equipment, bits of computer, and what felt like reams of paper, which contained not only instructions on how to manually install the various operating systems we supported but also build sheets containing configuration and information on the customer's required software stack I was deploying.

Once built, the server was packed back into its box, put in the back of someone's car, and taken to a data center. From there the server was racked and cabled for both power and networking and then powered the server on – if everything was configured correctly, it would spring into life and be available on the network.

After some quick testing, it was back to the comfort of the office to complete the build steps and, finally, hand the server over to the customer for them to deploy their software on.

While this process was fine when there was one or two of these deployments, once in a blue moon, as things got busier, it quickly became unmanageable.

The next logical step was to have a build server that contained drive images for all the supported operating systems and base software stack configurations, with some custom scripts that ran when the server first booted to customize the base configuration and get it onto the network when the server was racked in the data center.

Enter virtualization

Once we started to move from provisioning bare metal servers for customers to virtualized servers, things got a lot easier – for a start, as you didn't have to physically connect RAM, CPUs, or hard drives to the servers, assuming the cluster you were building the server in had the resource available, it made quite a dramatic change to the deployment time and also resulted in less time in the build room and data center.

At this point, we had built up a collection of custom scripts that connected to both the virtualization hypervisors and virtual machines – these were shared between the team members in our subversion repository and documented in our internal wiki.

This was my first, extremely basic by today's standards, introduction to Infrastructure as Code.

Virtual machine configuration

The next logical steps were to add a remote configuration into the mix by using a tool, such as **Puppet** or **Chef**; we could deploy our virtual machines using our custom scripts and then have the servers call back to our main management server, and then bootstrap itself as per the customer's desired configuration state.

Putting it all together

This was the final piece of the puzzle, which took our deployments from taking a few days per server to an hour or so, with the bulk of that time waiting for automated tasks to complete – though, as a lot of the initial stages of the deployments were initiated by our in-house DIY scripts, we still had to keep a careful eye on the progress.

This was because there wasn't much logic built in to handle errors or other unexpected hiccups during the deployment, which, in some cases, resulted in some challenging post-deployment problems – but the least said about those, the better.

Today's challenges

So, how do today's challenges differ from my own experiences? During my day job, I get to work with a lot of internal and external teams who are nearly all technical and are very hands-on with the day-to-day management and development of their own applications.

It's all documented

When discussing Infrastructure as Code with teams, one of the most common answers I get is as follows:

"We have the process to deploy our infrastructure documented, and anyone in the team can work through it to quickly deploy a resource."

While it is great that there is documentation and that it is accessible by all of the members of the team, you would be surprised, even with the presence of comprehensive and easy-to-follow documentation, at just how much variance there is when people come to actually implement it. They don't fully understand it because it is simply a set of tasks that lack any context as to why the tasks are being actioned.

Another potential issue is that the process is followed so often by a member of the team that they simply just get on with it, missing any updates or steps that have been added to the documentation.

Worst still – and this is more common than you may think – giving three technical people the same set of tasks to do can sometimes result in three very different outputs, as everyone has different experiences, which normally feeds into how we do things – for example, *last time I tried to A, B, and C, X happened, so now I do it C, B, and A* or *I think it would better to do it B, A, and then C – but don't have time at the moment to update the documentation*.

All of these can introduce inconsistencies in your deployments, which may go unnoticed as everyone thinks they are doing it correctly because they are all following the same set of documentation.

Next, next, next

The *next* (pun very much intended) answer I normally get is this:

"*We don't need to do it very often, and when we do, it's just clicking 'next, next, next' in an interface – anyone can do it.*"

When I get this answer, what I actually hear is, *The process to deploy the resource is so easy that we didn't bother to document it*. While this might be the case for some members of the team, not everyone may have the experience or confidence to simply click *next, next, next* to deploy and configure the resources without a guide.

As I am sure you can imagine, if it is possible for inconsistencies to be present when everyone is following the same set of documentation, then doing the deployment without any of the guardrails that the documentation puts in place is going to introduce even more potential issues further down the line.

Just because a resource has been deployed without error and works does not mean that it has been deployed securely and in such a way that could support your production workloads.

We have everything we need

The final most common answer when discussing Infrastructure as Code is as follows:

"*We have deployed everything we need and don't need any further resources.*"

Again, when I get this answer, I normally hear something slightly different – in my experience, this normally means that, a while ago, someone deployed something that is more than capable of the task and has now moved on, either going on to another project or department or has left the company altogether.

While it is great that the resources are running fine, this approach can cause issues if you ever need to redeploy or, worse still, firefight an issue with production, as a lot of knowledge of the underlying configuration is missing.

So, while you know *what's there*, you may not necessarily know *why*.

Conclusion

There are many more examples, but the previous ones are the most common ones I see when working with teams who may not have considered Infrastructure as Code to be a foundation of their deployment pipelines, and if you are reading this, then you may have already come across some of the examples and want to move onto the next step.

So why would you take an Infrastructure-as-Code approach to your deployments? Well, there are several reasons, which include the following:

- **Documentation**: While we have already mentioned documentation, it's important to note that if you employ Infrastructure as Code, your deployment is documented as part of your code as it defines the desired state of your infrastructure in a human-readable format.

- **Repeatable and consistent**: You should be able to pick up your code and deploy it repeatedly – sure, you may make some changes to things such as resource SKUs and names, but that should just be a case of updating some variables that are read at the time of execution rather than rewriting your entire code base.

- **Time-saving**: As I mentioned, in my own experience, it sometimes took days to deploy resources – eventually, that got down to hours and, with more modern cloud-based resources, minutes.

- **Secure**: Because you have your infrastructure defined in code, you know that you will have a well-documented end-to-end configuration ready to review as needed. Because it is easily deployable, you can quickly spin up an environment to review or deploy your latest fixes into, safe in the knowledge that it is consistent with your production configuration, as you are not relying on someone manually working through a step-by-step document where something may get missed or misinterpreted.

- **Cost savings**: I think you should never approach an Infrastructure-as-Code deployment with cost savings being at the top of the list of things you would like to achieve – but it is a most welcome nice-to-have. Depending on your approach, cost savings can be a byproduct of the preceding points. For example, do you need to run your development or testing infrastructure 24/7 when your developers may only need it for a few days a week at most?

  Well, that infrastructure can be deployed as part of your build pipeline with next to no or little effort. In that case, you may find yourself in the enviable position of only paying for the resources when you need them rather than paying for them to be available 24/7.

So, now that we have discussed my personal journey with Infrastructure as Code and also gotten an idea of the different scenarios where Infrastructure as Code may come in useful and the potential reasons why you would want to incorporate it into your day-to-day workflows, let's now discuss some of the basic concepts you need to know about before we start to talk about the tools we are going to look at for the remainder of the book.

What is meant by declarative and imperative?

In programming, there are different ways to give instructions to a computer to achieve the programmer's desired result. These ways of telling the computer what to do are known as **programming paradigms**.

In general, they refer to how we build programs from logical ideas such as `if` statements or loops. There are other classifications as well: functional, structured, object-oriented, and so on. Each of these describes a different kind of task that programmers might perform when writing code or thinking about code.

Imperative and **declarative** programming is the most fundamental way in which programmers think about defining their tasks and the two main ways in which we need to think about how we write and structure our Infrastructure as Code.

Before we discuss each way, let us define a quick Infrastructure-as-Code project.

Basic Infrastructure-as-Code project

The following diagram shows the basic infrastructure for deploying a single virtual machine in Microsoft Azure:

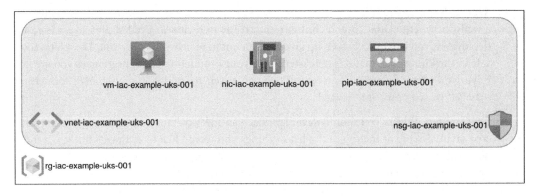

Figure 1.1 – The basic Infrastructure-as-Code project diagram

As you can see, the project is made up of the following components:

- **Resource group** (`rg-iac-example-uks-001`) – This is a logical container in Azure to store the resources.

- **Virtual network** (`vnet-iac-example-uks-001`) – This is a virtual network that will host our example virtual machine.

- **Subnet** (`snet-iac-example-uks-001`) – This is not shown in the diagram, but the virtual network contains a single subnet.

- **Network Security group** (`nsg-iac-example-uks-001`) – As we don't want management ports such as `3389` (RDP) or `22` (SSH) open to anyone on the internet, this will add some basic rules to only accept traffic on these ports from trusted sources. This will be attached to the subnet, so the rules apply to all resources deployed there.

- **Virtual machine** (`vm-iac-example-uks-001`) – This, as you may have guessed, is the virtual machine itself.

- **Network interface** (`nic-iac-example-uks-001`) – Here, we have the network interface, which will be attached to the virtual machine and the subnet within the virtual network.

- **Public IP address** (`pip-iac-example-uks-001`) – Finally, we have the public IP address; this is going to be attached to the network interface, which will allow us to route to the virtual machine from the trusted locations defined in the network security group.

While this is a basic infrastructure example, there are quite a few different resources involved in the deployment. Also, as we are going to be talking at a very high level about how this could be deployed, we won't be going into too much detail on Azure itself just yet, as this will be covered in *Chapter 4, Deploying to Microsoft Azure*.

Declarative approach

When talking about my own experiences, I mentioned that I used a configuration tool; in my case, this was Puppet. Puppet uses declarative language to define the target configuration – be it a software stack or infrastructure – but what does that mean?

Rather than try and give an explanation, let's jump straight in and describe how a declarative tool would deploy our infrastructure.

In its most basic configuration, a declarative tool only cares about the end state and not necessarily how it gets to that point. This means the tool, unless it is told to be, isn't resource-aware, meaning that when the tool is executed, it decides the order in which the resources are going to be deployed.

For our example, let us assume that the tool uses the following order to deploy our resources:

- Virtual network
- Resource group
- Network security group
- Subnet
- Public IP address
- Virtual machine
- Network interface

On the face of it, that doesn't look too bad; let us explore how this ordering affects the deployment of our resources in Azure.

The following figure shows the results of the deployments:

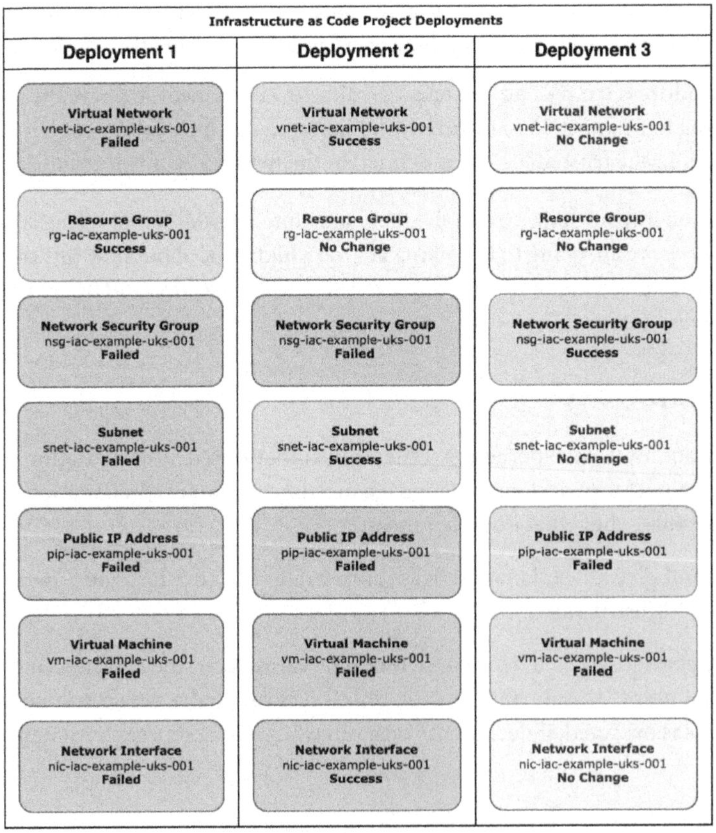

Infrastructure as Code Project Deployments		
Deployment 1	**Deployment 2**	**Deployment 3**
Virtual Network vnet-iac-example-uks-001 **Failed**	**Virtual Network** vnet-iac-example-uks-001 **Success**	**Virtual Network** vnet-iac-example-uks-001 **No Change**
Resource Group rg-iac-example-uks-001 **Success**	**Resource Group** rg-iac-example-uks-001 **No Change**	**Resource Group** rg-iac-example-uks-001 **No Change**
Network Security Group nsg-iac-example-uks-001 **Failed**	**Network Security Group** nsg-iac-example-uks-001 **Failed**	**Network Security Group** nsg-iac-example-uks-001 **Success**
Subnet snet-iac-example-uks-001 **Failed**	**Subnet** snet-iac-example-uks-001 **Success**	**Subnet** snet-iac-example-uks-001 **No Change**
Public IP Address pip-iac-example-uks-001 **Failed**	**Public IP Address** pip-iac-example-uks-001 **Failed**	**Public IP Address** pip-iac-example-uks-001 **Failed**
Virtual Machine vm-iac-example-uks-001 **Failed**	**Virtual Machine** vm-iac-example-uks-001 **Failed**	**Virtual Machine** vm-iac-example-uks-001 **Failed**
Network Interface nic-iac-example-uks-001 **Failed**	**Network Interface** nic-iac-example-uks-001 **Success**	**Network Interface** nic-iac-example-uks-001 **No Change**

Figure 1.2 – The results of deploying our infrastructure using a declarative tool

As you can see, it took three deployments for all the resources to be successfully deployed, so why was that?

- **Deployment 1**: The virtual network failed to be deployed as it needed to be placed with the resource group, which wasn't deployed yet. As all the remaining resources had a dependency on the virtual network, they also failed, meaning the only successful resource to be deployed during the first execution was the resource group, as that had no dependencies.

- **Deployment 2**: As we had the resource group in place from **Deployment 1**, then this time around, the virtual network and subnet both deployed; however, because the deployment of the network security group was attempted before the subnet was successfully deployed, that failed. The remaining failed resources – the public IP address and virtual machine – both failed because the network interface hadn't been created yet.

- **Deployment 3**: With the final set of dependencies in place from **Deployment 2**, the remaining resources – the network security group, public IP address, and virtual machine – all launched successfully, which finally left us with our desired end state.

The term for this is **eventual consistency**, as our desired end state is eventually deployed after several executions.

In some cases, the failures during the initial deployment of the resources don't really matter too much as our desired end state is eventually reached – however, with infrastructure, and depending on your target cloud environment, that may not always be true.

In the early days of Infrastructure as Code, this was quite a large issue as you had to build logic to consider dependencies for the resources you were deploying – which not only meant that you had to know what the dependencies were but the bigger your deployment, the more inefficient it became.

This is because the more logic you start adding to the code, the more you start working against the declarative nature of the tool, which also carries the risk of introducing race conditions when the code is executed. For example, if you have one resource that takes five minutes to deploy – how do you know it's ready? This would mean even more logic, which, if you got it wrong or something unexpected happened, you could be sat waiting for the execution to eventually time out.

Fear not; things have most definitely improved as the development of the tools has matured, and the tools have become more resource-aware. A lot of the manual logic you had to employ is now unnecessary, but there are still some considerations that we will go into in more detail in *Chapter 2, Ansible and Terraform beyond the Documentation*.

Imperative approach

As you may have already guessed, when using an imperative approach, the tasks execute in the order you define them – and we know the order in which we need to run the tasks to deploy our resources, which is as follows:

- Resource group

- Virtual network

- Subnet

- Network security group

- Network interface

- Public IP address

- Virtual machine

It means that the result of running our first deployment will look like the following:

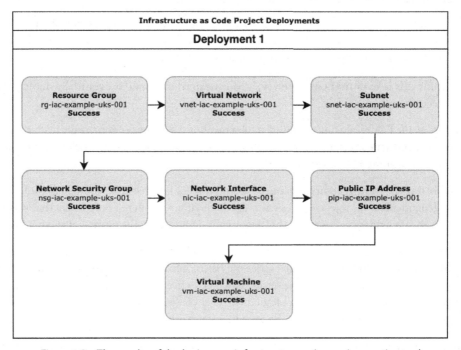

Figure 1.3 – The results of deploying our infrastructure using an imperative tool

Great, you may be thinking to yourself, it works the first time! Well, sort of; there is a big assumption that you know the order in which your resources need to be deployed, and you need to structure the code in such a way that takes that into account.

So, while this way would typically work first when executed, there could potentially be a little more upfront work to get the scripts in the right order using a little trial and error – however, once they are in the correct order, you can be confident that each time you execute them, they will work the first time.

Now that we have discussed the key differences between declarative and imperative when it comes to Infrastructure as Code, let's now talk about the differences between another deployment approach, pets versus cattle.

Pets versus cattle

Traditionally, *pets or cattle* has been a way of defining your data center resources. It's an analogy that describes a collection of hardware or virtualized resources as either pets or cattle.

Pets

Pets are resources that are owned by individual users/teams or managed on an individual basis.

Normally, they are seen as important fixed points within any application architecture and, like with a pet, you do the following:

- **You give them names**: For example, your server may have a hostname that looks something like `backendapplication.server.domain.com`, so it is easily identifiable.

- **You feed and water them**: For example, you take and keep backups that you review regularly. You keep a close eye on resource utilization and add more RAM and drive space as required.

- **If they get ill, you care for them**: They have monitoring agents installed, meaning you are alerted if there is a problem – sometimes 24/7 – and if there is an issue, you do everything you can to restore service by having troubleshooting procedures in place.

- **You expect them to live for a long time**: Given their importance within your application architecture and that you are caring for them, you expect them to be around for quite a while.

Resources that are now considered pets have typically been around for some time, and their configuration has organically evolved over that time based on their utilization, making them each a unique deployment, which is why you care for them, just like having a real pet – a good example of this is a long-running server.

Cattle

With resources that have been deployed to be treated as cattle, you only care about the health of the herd and not an individual resource:

- **There are too many of them to give names**: For example, your servers may have a hostname that looks something like `beapp001.server.domain.com` to `beapp015.server.domain.com`; you just keep incrementing the number rather than assigning a unique name that makes them easily identifiable.

- **You watch them from afar**: Given the number of resources, you only really care about the availability of the herd, meaning you probably only just ship performance stats and logs from the resources, and you do not need to back them up as it would be quick to replace them.

- **If they get ill, you replace them**: As already mentioned, if there is an issue with a resource, rather than troubleshoot the problem, you terminate it and replace it with another resource ASAP. Typically, this process is automated so that a resource is quickly taken out of service and another one put in place.

- **You don't expect them to live for a long time**: Given their numbers, they can be quite short-lived – in some cases, resources may only exist for a short amount of time to handle an increase in the workload. Once the demand for additional resources has ceased, some of the resources are terminated.

Conclusion

Pets versus cattle mainly applies to application deployment strategies rather than purely just the underlying infrastructure. After all, let's say your application, for whatever reason, needs to run as a single fixed point – for example, your application does the following:

- Writes essential files to the local disc, which can't be lost if an instance is terminated

- Has manual steps for bringing an application instance online after it has been deployed

- Is licensed to a MAC address or CPU ID of a host

In this case, you may not be able to treat your deployments as cattle, but you can write your Infrastructure as Code so that the bulk of your deployment is as automated as possible.

These are technical reasons, but there are some considerations from a business point of view as well.

The one that will get most businesses' attention is cost efficiency. Your choice of either a pets or cattle approach could have a significant impact on your hosting costs.

The cattle approach, which treats servers as ephemeral resources, allows for better resource utilization and automated scaling, potentially reducing costs. On the other hand, deploying pets, which emphasizes individual server care, may result in higher maintenance and management costs but could be justified for mission-critical applications that demand special attention.

Taking a cattle approach enables faster deployment and scaling of your workloads; this allows businesses to respond more quickly to market changes and customer needs. Deploying pets might lead to longer deployment times, potentially impacting a company's competitiveness.

Regulatory and security requirements could also influence the choice between pet and cattle deployments. The pets approach, focusing on managing individual resources, may be more appropriate for businesses with strict regulatory or security requirements, as it allows for more fine-grained control and auditing of server configurations. However, the cattle approach, emphasizing automation and rapid scaling, might not provide the same level of control and may require additional efforts to ensure compliance and security.

Now that we have a good idea of the type of deployments you could be dealing with, let's now talk about what this means for an Infrastructure-as-Code deployment.

What does all this mean for our Infrastructure-as-Code deployments?

So far, we have spoken a lot about some of the approaches and journeys people take to get to the point where they are considering using Infrastructure as Code, so before we look at some of the toolings in *Chapter 2, Ansible and Terraform beyond the Documentation*, let's talk about some of the actual use cases.

In my opinion, the most significant advantage of using Infrastructure as Code is **consistency** – if you need to repeat a process or deployment more than once, then define your deployment as Infrastructure as Code.

This will make sure that resources are deployed the same every time, no matter who is deploying them; if everyone is using the same set of code, then it stands to reason that the outputs will be the same (apart from variables you allow to override the values on such as SKUs, resource names, etc.).

An Infrastructure-as-Code approach not only gives you consistency between team members deploying the code but also between environments. Before I started defining my deployments as Infrastructure as Code, configuration drift between environments was quite a significant issue – environments were online for so long that *tweaks* were being applied and not carried through, so when code moved between my development, test, and finally, production environments, unexpected things would start to happen.

Next up is **collaboration**; as your infrastructure is defined in code, you can use the same development workflows you use for your applications. I am sure that most of you use a version control system for your code, more than likely Git via hosted services such as GitHub, GitLab, BitBucket, or Azure DevOps – if so, you have everything in place to track changes and collaborate on your infrastructure configuration.

You can also extend this further by introducing branching and pull requests based on your existing procedures to encourage change and testing, making the ongoing maintenance and development of your Infrastructure-as-Code projects genuinely **collaborative**.

Once you have your Infrastructure as Code hosted in version control, you can also take advantage of **automation**, again using the same processes and pipelines you use to build your application – using services such as GitHub Actions or Azure DevOps Pipelines.

Using services such as these gives you the ability to execute tasks from a single location that is covered by the service's role-based access control, rather than being reliant on each member of the team downloading and running the Infrastructure-as-Code deployments locally.

If a team member would be running it locally, then that would mean that each team member who needs access to deploy would also need quite a high level of access to target resources – such as the public cloud you are deploying to.

Using **automation** solutions such as the ones mentioned previously means that you can allow people to use credentials in their pipelines without them having to know what the credentials are. This means you can grant the individuals a lower level of access to your resources – such as *read-only* – as they only need to view resources rather than manage them.

One significant side effect of this approach is that because people don't have a level of access outside of the automation, they won't be tempted to *quickly jump into the portal and make a change to fix something manually* and instead will need to update the code and do a deployment, meaning that the change is tracked and the execution logged, so you know who did what, when, and why.

Finally, something that we have already mentioned – **cost savings**. If you have your Infrastructure-as-Code deployments in version control and automated, then it's not a stretch to deploy your infrastructure as needed rather than running it 24/7.

For example, if you have a pipeline to build your application, once that pipeline has successfully executed, then it can trigger, which builds the infrastructure – once built, that in turn triggers a deployment, and from there, your tests can run against the deployment and freshly deployed resources. The results of the test can be stored, and the infrastructure is then torn down as it is no longer needed.

This end-to-end process may take half an hour – but that's that half an hour's worth of resource cost versus paying for 24/7 resource costs – which I am sure you will agree is quite a saving.

Summary

In this chapter, we discussed and covered some of the core concepts we will be following throughout the remainder of the book. We talked about my own journey with Infrastructure as Code, which we will be picking up in further chapters.

We discussed some of the common questions that get raised when discussing Infrastructure-as-Code projects, along with some of the positive and negative feedback you may get. Then we went on to talk about the differences between the two deployment approaches.

The first is **declarative** and **imperative**, which is how your deployment code is executed and in which order.

The second approach we discussed, pets versus cattle, while not strictly an Infrastructure-as-Code method, does have relevance to the approach you would take to writing your Infrastructure-as-Code scripts.

As we get more hands-on, I will share some of my own challenges and successes with Infrastructure as Code.

Speaking of getting more hands-on, in our next chapter, *Chapter 2, Ansible and Terraform beyond the Documentation*, we are going to look at two of the most common Infrastructure-as-Code tools and start looking at some actual Infrastructure-as-Code examples, as well as get an idea of how concepts such as **declarative** and **imperative** apply to them. Plus, we will be covering some tips and tricks based on my own experience with the two tools.

Further reading

Here are links to more information on some of the topics, tools, and services that we have covered in this chapter:

- Puppet: `https://www.puppet.com/`
- Chef: `https://www.chef.io/`
- Microsoft Azure: `https://azure.microsoft.com/`
- GitHub: `https://github.com/about`
- GitLab: `https://about.gitlab.com`
- BitBucket: `https://bitbucket.org/product`
- Azure DevOps Repos: `https://azure.microsoft.com/en-us/products/devops/repos/`
- GitHub Actions: `https://github.com/features/actions`
- Azure DevOps Pipelines: `https://protect-eu.mimecast.com/s/8fwmCjvkghnR9PJIRuMq_?domain=azure.microsoft.com/`

2

Ansible and Terraform beyond the Documentation

The next phase in our journey to **Infrastructure as Code (IaC)** is to take a look at **Terraform**, an IaC tool from HashiCorp, and **Ansible**, an IaC and configuration management tool from *Red Hat*.

We will also compare the advantages and disadvantages of using them, set them up on macOS, Windows 11, and Ubuntu Linux, and look at using Visual Studio Code as an **integrated development environment (IDE)** to write our code, plus look at which recommended extensions to install.

In this chapter, we are going to take a look at the following topics:

- What is important when choosing a tool?
- Introducing Terraform
- Introducing Ansible
- Introducing Visual Studio Code, the open source IDE from Microsoft

Before we start looking at the tools we will be using throughout this title, let's quickly discuss a checklist I use to choose which tools to use in a project.

What is important when choosing a tool?

So, you have a new project – you know which cloud provider you will use, and your development team has given you an overview of their application – meaning you already have a good idea of the resources you will deploy and manage. You have been given free rein to choose which IaC tool to use – so how do you choose?

Personally, my approach is always to use the best tool for the job rather than trying to fit the job to the tool – that, in my experience, always ends up causing issues when it comes to deploying the code and managing the deployment once it has been deployed.

Let us discuss some of the key things you will need to consider.

Deployment types

There are two main types of deployment I come across, with the first being using IaC to repeatedly deploy the same resources in a predictable and consistent way.

The most common use case for this approach is for dev, test, and other lower environments, not production.

The goal is to integrate with your developer's build, release, and test pipelines so that when they push their code changes for one of the environment branches mentioned, the following happens:

- The push triggers the deployment of the resources using your IaC scripts

- Once the resources have been deployed, your IaC pipeline hands back over to developers' pipelines for them to build their code and deploy it to the resources that have just been launched

- Once the application code has been deployed, run the developer's automated testing or notify someone within the team that the newly pushed code is ready for manual testing

- Finally, with testing complete, after the results are stored and either by an automated or manual decision gate in the pipeline, the resources deployed at the start of the process are terminated

The process above is repeated for each push – with multiple deployments sometimes being executed in parallel.

The advantage of taking an approach like this is that you not only get cost savings, but you are only running resources while they are needed. Additionally, configuration drift is avoided as you are launching the resources from scratch with each deployment.

Configuration drift occurs when someone with the best intentions quickly tweaks a setting somewhere manually to get something working and doesn't document it anywhere. In this case, ad-hoc fixes would need to be rolled into your code for them to persist through to the next deployment.

The next deployment type is to launch and manage your resources using your IaC scripts. As you may have guessed, this approach is used in longer-running environments such as production.

When you first think about this type of deployment, it is easy to assume that it is quite close to the first deployment type – however, in fact, the first deployment type is only executed once per deployment, whereas this type is executed multiple times against the same deployment which can introduce some interesting challenges, such as the following:

- Depending on the resource type, where is the line drawn between what is configured and managed by your IaC scripts and the application deployment?

- As you are dealing with long-running resources, what additional logic or error checking do you need to build into your IaC scripts so that the execution of your code is terminated rather than the resources you are running the code against? After all, you don't want to cause an outage, no matter how easy it is to recreate your infrastructure!

- How are you managing the state of your infrastructure? As we will learn in the next section of this chapter, having a consistent state is important for one of the tools we will cover in this book – so where is it stored long term?

Infrastructure and configuration

While we will be talking a lot about IaC in this book, which I am hoping if you have made it this far shouldn't be a surprise, where is the line drawn between your IaC scripts and the deployment/ configuration of the application?

A good example of this is when your project involves deploying **Infrastructure as a Service (IaaS)** resources such as virtual machines. Let's say you need to deploy two Linux servers and then install NGINX along with a scripting language such as PHP on there; how will you achieve that?

Most public cloud providers allow you to attach and execute a script when launching a virtual machine using a service such as cloud-init – while this should cover most basic use cases, using this approach does add a level of abstraction that could cause problems – for example, does your cloud provider provide any details on the execution of the script – and will your IaC execution know if that has failed?

If you need more granular control or visibility of the commands being run as part of the deployment, then this will dictate which tool you choose as a pure IaC tool may not be enough for your needs.

This will also influence the next decision.

External interactions and secrets

As mentioned at the end of the last section, if your IaC script needs to interact with a resource using a service outside of a publicly accessible API – such as **Secure Shell (SSH)** or **Windows Remote Management (WinRM)** to run scripts on a virtual machine or an internally hosted API such as the vSphere API used to manage resources hosted in VMWare environments, then you will carefully need to choose where your IaC is executed from as you will need line of sight to the resource you are interacting with.

Likewise, depending on how you manage secrets within your IaC scripts for things such as passwords or certificates for services you are launching, you will also need line of sight, by which I mean direct access, of your secret storage or a way of securely injecting them into your scripts because as storing them as *hardcoded values in plain text within your IaC is never an option, ever!*

This means that you will need to assess where and how you execute your scripts, considering things such as firewalls plus access to resources and credentials – all without exposing any secrets.

We will cover all of this when we roll up our sleeves and start building up our deployment in later chapters.

Ease of use

The final consideration is simply how easy the tool is to use.

It is easy to get swept up in the latest shiny new technologies, but if you are the only person in your team who has any experience with it, you will be adding complications as not only will you need to up-skill the rest of the team so they can also work with the code but also you will need to deal the issues that can occur from being an early adopter.

Summary

Everything we have discussed in this section should be at the forefront of your mind when approaching any IaC project. By the end of this book, you will have both the answers and experience to all of the questions and considerations raised during this section to be able to choose the right tool for the job rather than trying to fit your project to the tool, or what sometimes can only be described as fitting a square peg in a round hole.

Now is the time I hope you have been waiting for; we will look at our two main tools.

Introducing Terraform

The first of the two tools we will look at is Terraform by HashiCorp.

Hashicorp Terraform is an enterprise-ready cloud and virtualization management tool. It helps you manage your resources and deploy new instances with ease. Terraform is an open source tool for managing the cloud infrastructure, allowing you to not only efficiently configure and deploy your resources but also help you maintain your infrastructure while evolving it over time.

Terraform has a unique architecture in that it uses a state machine to manage resources and it is fully modular, and you can scale the service as per your needs. Finally, it is also integrated with many third-party tools and services.

Terraform uses the **Hashicorp Configuration Language** (**HCL**). You could be mistaken at first glance for thinking it is for JSON or YAML, but it's a syntax and API designed by HashiCorp for building structured configuration formats, whereas YAML and JSON are just formats that define data structures in human- and machine-readable formats respectively.

Rather than going into any more detail about HCL – let's take a look at an HCL example.

An HCL example – creating a resource group

I personally do a lot of my day-to-day work with Microsoft Azure, so I will target that with this example.

> **Information**
>
> Feel free to follow along; if you need assistance installing Terraform, then there are links to the relevant documentation in the *Further reading* section at the end of this chapter.

Azure has a concept of resource groups that act as a logical container for your resources, so let us start by creating one of those:

1. There are three main sections we need in our Terraform, the first of which tells Terraform which version of Terraform our code is compatible with and also which external providers we need to use. In the case of creating a resource group, this looks like the following:

    ```
    terraform {
      required_version = ">=1.0"
      required_providers {
        azurerm = {
          source  = "hashicorp/azurerm"
          version = "~>3.0"
        }
      }
    }
    ```

    One of the biggest selling points of Terraform is that it is both machine- and human-readable – I am sure you will agree from the small preceding snippet of code that it is easy to figure out what is going on.

    Here we have said that `required_version` of Terraform should be greater or equal to `1.0`. Next up, we have `required_providers`; a provider is an external library that extends the functionality – in this example, we are telling Terraform to download and use the latest version of the `3.0` release of the `azurerm` provider from `hashicorp/azurerm`, which is where the official provider releases should be sourced from.

2. The next section configures the providers. For our example, we won't do any additional configuration, so this just looks like the following:

    ```
    provider "azurerm" {
      features {}
    }
    ```

3. Next up is the final section of our example; this is where we configure our resource group:

    ```
    resource "azurerm_resource_group" "example" {
      name    = "rg-example-uks"
    ```

```
    location = "UK South"
}
```

As you can see, there is not much to it – we simply define what we want the resource to be called by providing `name` and also which Azure region we would like the resource group to be placed in using `location`.

All the preceding code is placed in a file called `terraform.tf` in an empty folder. Before we can create the resource group, we will need to initialize Terraform; this will download the `azurerm` provider and create a few supporting files, such as `locks`, which are needed to execute the code.

4. To deploy the resource group, we first need to run the following command to prepare our local environment:

    ```
    $ terraform init
    ```

5. This will give something like the following output:

    ```
    Initializing the backend...
    Initializing provider plugins...
    - Finding hashicorp/azurerm versions matching "~> 3.0"...
    - Installing hashicorp/azurerm v3.32.0...
    - Installed hashicorp/azurerm v3.32.0 (signed by HashiCorp)
    Terraform has created a lock file .terraform.lock.hcl to record
    the provider selections it made above. Include this file in
    your version control repository so that Terraform can guarantee
    to make the same selections by default when you run "terraform
    init" in the future.
    Terraform has been successfully initialized!
    ```

6. So now that Terraform is ready, we can run it – first of all, we need to run a plan:

    ```
    $ terraform plan
    ```

    This should give us an idea of what Terraform is going to do when we apply our configuration; in my case, this gave the following output:

    ```
    Terraform used the selected providers to generate the following
    execution plan. Resource actions are indicated with the
    following symbols:
      + create
    Terraform will perform the following actions:
      # azurerm_resource_group.example will be created
      + resource "azurerm_resource_group" "example" {
          + id       = (known after apply)
          + location = "uksouth"
          + name     = "rg-example-uks"
        }
    Plan: 1 to add, 0 to change, 0 to destroy.
    ```

What Terraform has done here are some basic flight checks, discovering that it doesn't know about a resource group called `rg-example-uks` in the `uksouth` region and, therefore, it needs to add it, and because we are only creating a single resource there is 1 to add.

7. To create the resource group, we need to run the following command:

    ```
    $ terraform apply
    ```

    When doing so, it will give us the same output as when running `terraform plan`, but this time, as usual, if we want to proceed, answering `yes` will then deploy the resource:

    ```
    Do you want to perform these actions?
      Terraform will perform the actions described above.
      Only 'yes' will be accepted to approve.
      Enter a value: yes
    azurerm_resource_group.example: Creating...
    azurerm_resource_group.example: Creation complete after 0s
    [id=/subscriptions/xxxxxxxx-xxxx-xxxx-xxxx-xxxxxxxxxxxx/
    resourceGroups/rg-example-uks]
    ```

8. There we have it; our resource group has been created. Running the `terraform apply` command again gives the following output:

    ```
    azurerm_resource_group.example: Refreshing state... [id=/
    subscriptions/xxxxxxxx-xxxx-xxxx-xxxx-xxxxxxxxxxxx /
    resourceGroups/rg-example-uks]
    No changes. Your infrastructure matches the configuration.
    Terraform has compared your real infrastructure against your
    configuration and found no differences, so no changes are
    needed.
    Apply complete! Resources: 0 added, 0 changed, 0 destroyed.
    ```

So, no changes are needed – now let's add another resource – how about a storage account?

Adding more resources

Follow these steps to add a storage account:

1. To do this, we simply need the following resource at the end of the `terraform.tf` file:

    ```
    resource "azurerm_storage_account" "example" {
      name                     = "saiacforbeg2022111534"
      resource_group_name      = "rg-example-uks"
      location                 = "UK South"
      account_tier             = "Standard"
      account_replication_type = "GRS"
    }
    ```

Running `terraform apply` now gives the following output, which I have truncated as the total amount of lines has gone from 13 to 166 lines:

```
azurerm_resource_group.example: Refreshing state... [id=/
subscriptions/xxxxxxxx-xxxx-xxxx-xxxx-xxxxxxxxxxxx /
resourceGroups/rg-example-uks]

Terraform used the selected providers to generate the following
execution plan. Resource actions are indicated with the
following symbols:
  + create
Terraform will perform the following actions:
  # azurerm_storage_account.example will be created
  + resource "azurerm_storage_account" "example" {
      + account_kind                = "StorageV2"
      + account_replication_type    = "GRS"
      + account_tier                = "Standard"
      + location                    = "uksouth"
      + name                        =
"saiacforbeg2022111534"
      + resource_group_name         = "rg-example-uks

Plan: 1 to add, 0 to change, 0 to destroy.
Do you want to perform these actions?
```

2. Answer `yes` to get the following output:

```
azurerm_storage_account.example: Creating...
azurerm_storage_account.example: Still creating... [10s elapsed]
azurerm_storage_account.example: Still creating... [20s elapsed]
azurerm_storage_account.example: Creation complete after 25s
[id=/subscriptions/xxxxxxxx-xxxx-xxxx-xxxx-xxxxxxxxxxxx /
resourceGroups/rg-example-uks/providers/Microsoft.Storage/
storageAccounts/saiacforbeg2022111534]
```

3. So, we now have our storage account – great, let's destroy it and run it again:

```
$ terraform destroy
```

The output of this command will tell us what is going to be removed (again, the output has been truncated):

```
Terraform used the selected providers to generate the following
execution plan. Resource actions are indicated with the
following symbols:
  - destroy
Terraform will perform the following actions:
  # azurerm_resource_group.example will be destroyed
  - resource "azurerm_resource_group" "example" {}
  # azurerm_storage_account.example will be destroyed
```

```
      - resource "azurerm_storage_account" "example" {}
    Plan: 0 to add, 0 to change, 2 to destroy.
    Do you really want to destroy all resources?
```

4. Answer yes, and that will, as you may have guessed, destroy the resources:

```
azurerm_resource_group.example: Destroying... [id=/
subscriptions/xxxxxxxx-xxxx-xxxx-xxxx-xxxxxxxxxxxx /
resourceGroups/rg-example-uks]
azurerm_storage_account.example: Destroying... [id=/
subscriptions/xxxxxxxx-xxxx-xxxx-xxxx-xxxxxxxxxxxx /
resourceGroups/rg-example-uks/providers/Microsoft.Storage/
storageAccounts/saiacforbeg2022111534]
azurerm_storage_account.example: Destruction complete after 3s
azurerm_resource_group.example: Destruction complete after 46s
Destroy complete! Resources: 2 destroyed.
```

5. Now running the script again using terraform apply tells us that two resources are going to be added:

```
Plan: 2 to add, 0 to change, 0 to destroy.
```

6. However, when you say yes and try to proceed, it will give an error:

```
| Error: creating Azure Storage Account "saiacforbeg2022111534":
storage.AccountsClient#Create: Failure sending request:
StatusCode=404 -- Original Error: Code="ResourceGroupNotFound"
Message="Resource group 'rg-example-uks' could not be found."
|    with azurerm_storage_account.example,
|    on terraform.tf line 21, in resource "azurerm_storage_
account" "example":
|    21: resource "azurerm_storage_account" "example" {
```

Why did it error? Let's take a look at the error and figure out what happened.

Fixing the error

First of all, why are we getting the error?

If you remember, in the previous chapter, we discussed the differences between imperative and declarative; this is an example of what happens if you don't plan your deployment right with an imperative tool.

As the storage account is attached to a resource group, and at the time of execution, the resource group didn't exist, and the storage account couldn't be created.

However, the resource group had no dependency failures as part of the Terraform run, meaning if you were to run terraform apply again, the storage account would be created – so how do we get around this so that it works the first time we run terraform apply?

You may have noticed that Terraform refers to the two resources we are creating as `azurerm_resource_group.example` and `azurerm_storage_account.example`; these are internal references that we can use in our own code. Also, for most of these references, some outputs are only populated once the resource has been created. Some of these references are only known after the resource has been created because it is a return value of the resource being created in Azure, such as a unique ID, while for others, they are ones which we have defined – but are only populated once the resource has been launched. In the case of `azurerm_resource_group`, the name and location are populated as an output value once the group has been created.

We can reference these in our `azurerm_storage_account` block by referring to the resource; this looks like the following:

```
resource "azurerm_storage_account" "example" {
  name                     = "saiacforbeg2022111534"
  resource_group_name      = azurerm_resource_group.example.name
  location                 = azurerm_resource_group.example.location
  account_tier             = "Standard"
  account_replication_type = "GRS"
}
```

What this will do is wait for the resource group to be deployed before Terraform will attempt to create the storage account – rather than just attempting to create both resources at the point of execution and failing.

While I wouldn't describe scenarios like this as errors or faults, they are more like quirks that you won't discover until you attempt something. Because of this, as we progress through the book, I will call out quite a few examples of this and other similar approaches because the more complex your deployment code is, the more considerations you will need to make when writing it.

The following screenshot shows the resources deployed in the Azure portal:

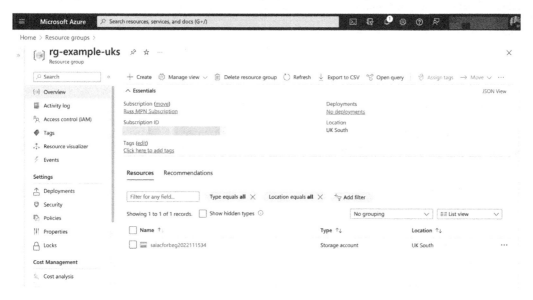

Figure 2.1 – The deployed resources within the Azure portal

You can clean up the resources you have launched by running the following command:

```
$ terraform destroy
```

This permanently deletes the resource group and storage account, so please ensure that you are happy to proceed before saying Yes.

Now that we have learned a little about Terraform, let's look at the other tool we will be using in the book, Ansible.

Introducing Ansible

The second tool we cover in detail in this book is Ansible by Red Hat.

Ansible is a popular configuration management tool that enables users to automate the deployment and management of their applications.

It uses a hub-and-spoke model where a controlling machine instructs other machines to perform tasks. You can use it to manage your servers, deploy applications, or configure your network devices. One of the biggest advantages over other agentless devices is you don't need to install anything on the target device you're managing.

It supports YAML and JSON for writing playbooks, the main configuration file, meaning that it is language-agnostic when managing your remote systems and their state.

There is no one-size-fits-all solution when it comes to your IaC solution, and Ansible allows you to choose from various modules to achieve your desired result, providing a great deal of flexibility when managing your infrastructure.

An Ansible example

Let's take the same example we used for Terraform and recreate it in Ansible, creating an Azure resource group and placing an Azure storage account in it:

> **Information**
>
> Again, feel free to follow along; if you need assistance installing Ansible, there are links provided in the *Further reading* section at the end of this chapter.

1. Place the following code in a blank file on your local machine called `playbook.yml`:

```yaml
---
- name: Ansible Infrastructure as Code example
  hosts: localhost
  tasks:
    - name: Create an example resource group
      azure.azcollection.azure_rm_resourcegroup:
        name: "rg-example-uks"
        location: "UK South"

    - name: Create an example storage account
      azure.azcollection.azure_rm_storageaccount:
        resource_group: "rg-example-uks"
        name: "saiacforbeg2022111534"
        account_type: "Standard_GRS"
```

> **Hint**
>
> As this is a YAML file, the indentation is extremely important – before attempting to execute the playbook, I would recommend using an online tool such as `https://www.yamllint.com/` to quickly validate your file.

2. Once you are ready to run your playbook, you can run the following command:

```
$ ansible-playbook playbook.yml
```

On the first run, this gives the following error:

```
[WARNING]: No inventory was parsed, only implicit localhost is
available
[WARNING]: provided hosts list is empty, only localhost is
available. Note that the implicit localhost does not match 'all'
PLAY [Ansible Infrastructure as Code example] ******************
*************************************
TASK [Gathering Facts] *****************************************
*************************************
ok: [localhost]


TASK [Create an example resource group] ************************
*************************************
An exception occurred during task execution. To see the full
traceback, use -vvv. The error was: ModuleNotFoundError: No
module named 'msrest'
fatal: [localhost]: FAILED! => {"changed": false, "msg": "Failed
to import the required Python library (msrestazure) on Russs-
Laptop.local's Python /opt/homebrew/Cellar/ansible/6.6.0/
libexec/bin/python3.10. Please read the module documentation and
install it in the appropriate location. If the required library
is installed, but Ansible is using the wrong Python interpreter,
please consult the documentation on ansible_python_interpreter"}
PLAY RECAP *****************************************************
*************************************
localhost: ok=1    changed=0    unreachable=0    failed=1
    skipped=0    rescued=0    ignored=0
```

The first two warnings can be ignored; however, the error is something we will need to take care of before we can run our playbook.

3. As mentioned in the introduction of this section, Ansible is modular – these modules are known as collections. As you can see from the code, we are using the `azure.azcollection` collection.

    To install it, we need to run two commands; the first downloads the collection itself, and the second installs the required Python dependencies needed for the collection to work:

    ```
    $ ansible-galaxy collection install azure.azcollection
    $ pip3 install -r ~/.ansible/collections/ansible_collections/
    azure/azcollection/requirements-azure.txt
    ```

4. Once installed, rerun the following command:

    ```
    $ ansible-playbook playbook.yml
    ```

This should result in the following output (I have removed the warnings this time; as already mentioned, they can be ignored for now):

```
PLAY [Ansible Infrastructure as Code example] ******************
*********************************************
TASK [Gathering Facts] *****************************************
*********************************************
```

```
ok: [localhost]
TASK [Create an example resource group] ************************
********************************************
changed: [localhost]
TASK [Create an example storage account] **********************
********************************************
changed: [localhost]
PLAY RECAP ****************************************************
********************************************
localhost: ok=3    changed=2    unreachable=0    failed=0
        skipped=0    rescued=0    ignored=0
```

As you can see, everything went as planned this time, and of the three tasks executed (the first being a check on localhost), two show changes.

5. Running the command again results in a play recap showing three OKs:

```
PLAY RECAP *****************************************************************
********************************************
localhost: ok=3    changed=0    unreachable=0    failed=0
        skipped=0    rescued=0    ignored=0
```

You may have also noticed that it just ran the first time – if you ignore installing the prerequisites.

Unlike Terraform, Ansible, when executed this way, is declarative. This means that it ran the tasks in order and waited for each of them to complete before progressing to the next task in the playbook file. This means that we didn't find ourselves in a situation where Ansible was trying to launch resources linked to other resources that don't already exist.

Another key difference between Ansible and Terraform is that Ansible is stateless – which means Ansible does not track or store the state of your resources in a file and instead looks at each resource at the time of execution.

Personally, I think this is one of the critical differences between Terraform and Ansible, as I have lost count of the times that I have had to debug problems because someone or something has made a change to a resource outside of Terraform, which Terraform has then struggled to reconcile between the resources that are actually there and ones that it thinks are there.

Finding yourself in this situation is dangerous territory if you don't pay attention.

You may find that the only way Terraform can get its state back to how it thinks it should be deployed is to start terminating and redeploying resources – which would cause all sorts of chaos if you were in a production environment.

On the other hand, because Ansible does not keep track of the state of the resources it manages, it won't know about a resource's state or configuration until you execute the playbook.

99% of the time, running an Ansible playbook will execute tasks that launch or update existing resources, so Ansible not keeping track of the state is not an issue – in fact, it could be a benefit as it is not trying to enforce a state it knows about.

The one downside to this is that because it doesn't know what resources are present, there is not an Ansible equivalent of the `terraform destroy` command. When you run this command in Terraform, it simply removes the resources present within the state file giving a convenient way of removing everything Terraform is managing.

To get around this with Ansible, I normally provide a second playbook that sets the state of all or just some of the resources to `absent` – given the default state for most resources is `present`, this will remove the resources listed.

In the example we have just covered, the playbook to remove the resources looks like the following:

```
---
- name: Ansible Infrastructure as Code example
  hosts: localhost

  tasks:
    - name: Terminate the example resource group
      azure.azcollection.azure_rm_resourcegroup:
        name: "rg-example-uks"
        location: "UK South"
        state: absent
        force_delete_nonempty: true
```

You may have noticed an empty line (- - -) at the end of the code block; it is important that these are present.

> **Warning**
>
> You may have noticed that we are setting the `force_delete_nonempty` flag to `true` in the preceding code snippet. Please be careful when using that flag as you will not be asked if you are sure, and this overrides the default action of failing because there are resources within the resource group.

Place the preceding code in a file called `destroy.yml` and run the following command:

```
$ ansible-playbook destroy.yml
```

This will delete the resource group. Because the storage account is a child resource within the resource group and we have instructed Ansible to remove resource groups even if they are not empty, it will also be removed.

Now that we have learned about Ansible, let's look at a tool we can use to write our code.

Introducing Visual Studio Code

The final tool I am going to introduce isn't an IaC tool but an IDE, which is used to write the code itself.

Visual Studio Code is a powerful code editor perfect for most development languages, including your IaC projects. It is feature-rich, fast, and highly customizable, making it the ideal choice no matter which of the tools you decide to go with.

The best part is that Visual Studio Code is completely free and open source. Whether you're a professional web developer, system administrator, or DevOps practitioner, Visual Studio Code has everything you need to create well-structured code.

Is it something I use on a daily basis – as you can see from the following screenshot, via the use of extensions, you get features such as syntax highlighting:

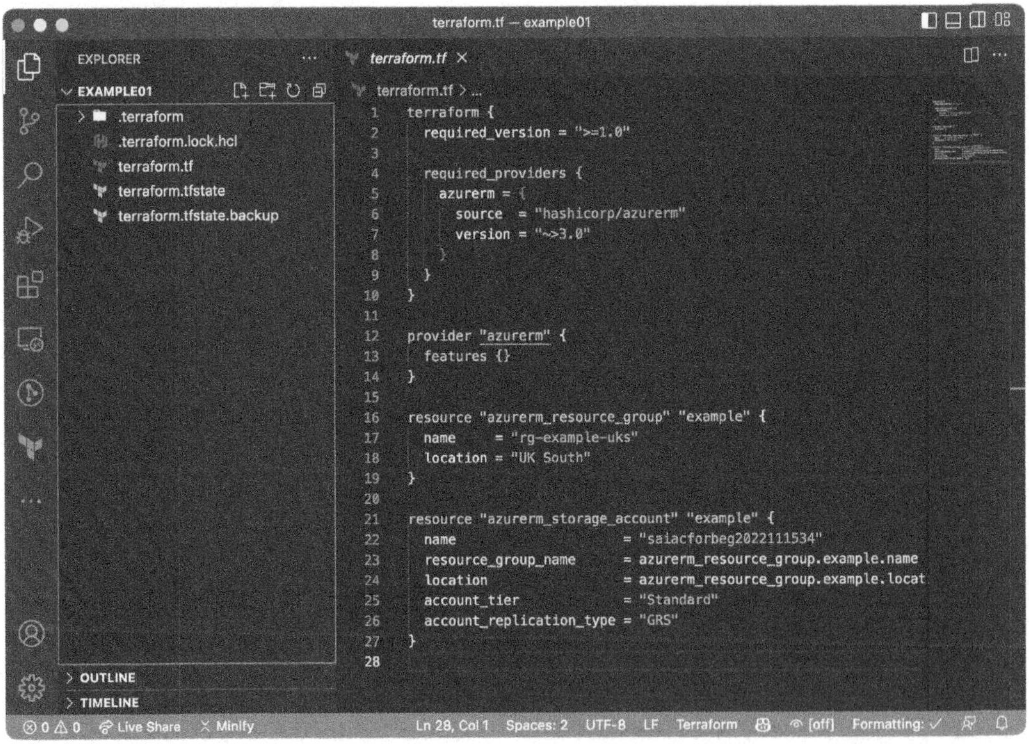

Figure 2.2 – Our Terraform example opened in Visual Studio Code

But beyond syntax highlighting, with extensions you can also get powerful features such as the following:

- **Inline error checking**: This is where your code is checked for syntax errors and general issues, such as referencing a variable or output that does not exist, and makes you aware of them

- **Auto-complete**: This functionality varies between extensions, but they can fill in details as you type, suggesting which flags/keywords and values could be used

- **Formatting**: As already mentioned, formatting is really important when it comes to both HCL and YAML; there are extensions for both languages that will check your formatting as you type and auto-correct if there are problems, which should hopefully save you from having to use an online tool such as the one linked in the Ansible section

- **Version Control and continuous integration/continuous delivery (CI/CD)**: There is a built-in integration with Git, as well as extensions for services such as GitHub, Azure DevOps, and other popular version control and CI/CD tools and services

While it is not essential to use an IDE such as Visual Studio Code, I think you will miss out on a lot of functionality and troubleshooting help if you don't.

For details on where to get Visual Studio Code from as well as recommended extensions that will be helpful throughout this book, please see the *Further reading* section.

Summary

In this chapter, we got a very quick feel for the approach and considerations when it comes to choosing the right IaC tool for your project.

We also looked at both Terraform and Ansible and some of the small differences between the two tools before discussing Visual Studio Code, which I hope you will install and make use of.

In the next chapter, which is the last chapter of *Part 1* of this book, we will look at the example project, which we will execute throughout the remainder of the book and get our teeth into both Terraform and Ansible on the two major public cloud providers.

Further reading

Here are some resources to help you delve deeper into Terraform:

- Main website: `https://www.terraform.io/`
- Download and install guides: `https://developer.hashicorp.com/terraform/downloads`
- Azure Resource Manager provider: `https://registry.terraform.io/providers/hashicorp/azurerm/latest`

Here are some resources to help you delve deeper into Ansible:

- Main website: `https://www.ansible.com/`
- Download and install guides: `https://docs.ansible.com/ansible/latest/installation_guide/index.html`
- Azure collection: `https://galaxy.ansible.com/azure/azcollection`

Here are some resources to help you delve deeper into Visual Studio Code:

- Main Website: `https://code.visualstudio.com/`
- Downloads: `https://code.visualstudio.com/Download`
- HashiCorp Terraform extension: `https://marketplace.visualstudio.com/items?itemName=HashiCorp.terraform`
- Red Hat Ansible extension: `https://marketplace.visualstudio.com/items?itemName=redhat.ansible`
- GitHub Repositories extension: `https://marketplace.visualstudio.com/items?itemName=GitHub.remotehub`

3

Planning the Deployment

In this chapter, we will dive into the important stage of planning the deployment of our **Infrastructure as Code** workload. Before we can deploy our infrastructure, it is crucial to have an understanding of what we are deploying and how we want to approach the deployment process. This will ensure that our deployment is efficient, streamlined, and free of errors.

We will start by introducing the workload that we will be deploying in the following two chapters. This will give us a clear understanding of what we are trying to achieve and what resources we need to deploy.

Next, we will discuss how to approach the deployment of our infrastructure. This will include a step-by-step guide on how to plan and execute the deployment process in a smooth and effective manner. We will also discuss best practices and tips for ensuring a successful deployment.

Finally, we will examine the high-level architecture of our infrastructure. This will give us an overview of how the various components of our infrastructure will fit together and interact with each other.

With this understanding of our workload and deployment approach, we can move forward confidently to *Chapter 4, Deploying to Microsoft Azure*, and *Chapter 5, Deploying to Amazon Web Services*, where we go into the low-level design and deployment code.

Planning the deployment of our workload

When coming up with sample workloads for projects, it is sometimes difficult to find something that isn't too complex, but also not so simple that the example is simply a case of following steps 1 through 10 and you are done. To ensure that the project we are going to be covering is both exciting and has the sort of considerations you will need to make in your projects, but also is something that most of you will have had some experience with at one point or another, I have chosen to use **WordPress**.

WordPress is an open source **content management system (CMS)**, hosted using PHP and MySQL, that enables you to build websites and blogs. It was developed in 2003 and has grown to be one of the most popular CMS platforms in the world, running millions of websites. WordPress is renowned for its simplicity and flexibility, which makes it a great choice for users of all skill levels. I can hear what you are thinking: *But WordPress has its famous five-minute installation, which is just a case of following a few simple steps!* However, in our case, we are going to look at deploying WordPress across multiple hosts as well as using services native to each of the public clouds we are going to be targeting for the database, storage, and networking layers.

One of the key features of WordPress is its use of themes, which allow users to easily change the look and feel of their website without having to modify the underlying code. This makes it simple for users to produce professional-looking websites without having any knowledge of web design or programming. In addition to WordPress having a large and active community of users and developers, there are always new features and updates being released. This, in combination with its open source nature, makes it an excellent choice for anybody looking to develop a website or blog.

Before we progress any further, a word of warning.

> **Information**
>
> While we will be deploying WordPress across multiple instances, this approach is being taken to give an example of the considerations you need to make tackling your own Infrastructure as Code projects; please do not use it as a guide for deploying and managing your own highly available WordPress installations.

So, now that we have an idea of what we are going to be deploying, let us look at answering some of the questions that you may have to give an idea of how we are going to be approaching the deployment.

How to approach the deployment of our infrastructure

First, as we have already mentioned, WordPress runs on top of PHP and MySQL; to be more explicit, it has the following requirements:

- A piece of web server software such as Apache or NGINX
- PHP version 7.4 or greater
- MySQL version 5.7 or greater, or MariaDB version 10.3 or greater

> **Information**
>
> Please note that, at the time of writing, PHP 8 only has beta support in WordPress version 6.1; because of this, we will be installing PHP 7 in our example deployment.

Deployment considerations

So, we know from our requirements that we are going to need a web server and PHP installed on something, while all the cloud services we are going to be looking at in the next two chapters offer some sort of application hosting as a service. For our project, we are going to use **virtual machine instances** running Ubuntu.

Rather than launching a single host, which would be a single point of failure, let's look at launching a minimum of two virtual machine instances to run WordPress on. This approach introduces some complexity as WordPress really likes to run as a fixed point, so what considerations do we need to make when running WordPress across more than a single virtual machine instance?

- Shared storage across our hosts – all the WordPress code and files should be stored on a filesystem that is available across all the virtual machine instances we are running. As we are running Ubuntu Linux, this should be NFS rather than Samba or a Windows File share – this should be a **Platform-as-a-Service** offering from the cloud provider.

- When installing WordPress using our Infrastructure as Code scripts, we should look at only doing this from a single virtual machine instance and only once – let's refer to this as our **admin host**. All other hosts, or web hosts, should have all the packages needed to run WordPress installed and configured, and then mount the NFS share once WordPress has been successfully bootstrapped.

- As well as needing a way of distributing traffic across our multiple virtual machine instances, we are also going to have to think about how we will be serving traffic for the WordPress administration section of our website.

What about the **database**? As both the cloud services we are going to be targeting offer MySQL as a service, we will be utilizing these with our deployments. *Great*, you may be thinking to yourself – yes, it is one less resource we need to manage on our virtual machine instances – but there are also some considerations we need to make here as well, quite big ones:

- We will need to know the endpoint of the database host and the credentials we need to access it before we do the initial bootstrap of WordPress.

- We would want to lock down our database endpoint to only our virtual machine instances, as these will be the only things that will need access to it.

- We should also set up database backups!

Like the database-as-a-service and as already mentioned, we should be using an as-a-service for our **shared storage** running NFS; there are some considerations to make there too:

- We need to know the NFS endpoints so they can be mounted before we bootstrap WordPress, as we are going to need WordPress to be correctly installed before we launch the additional virtual machine instances

- Again, like the database, the NFS service needs to be locked down to only allow trusted virtual machine instances to be able to connect to it – we don't want just anyone to be able to connect randomly and be able to browse/download the contents of our filesystem

There are other aspects we also need to think about in our deployment:

- **Private networking**: As we want to lock things down, we will need some sort of internal network to launch our resources into

- **Load balancing**: We need a Layer 7 load-balancing service to distribute traffic across our backend services

- **Bootstrapping**: We will need to bootstrap both the software stack and WordPress itself on the virtual machine instances

Now that we are aware of the key considerations, let's look at the specific tasks that we will need to perform for deployment.

Performing deployment tasks

Based on the information in the previous section, we have a rough idea of what needs to happen and in which order. This all starts with launching the resources in our preferred cloud provider. To deploy our workload using Infrastructure as Code, we will need to perform the following tasks:

1. Launch and configure resources required for our private network.

2. Launch and configure the database as a service.

3. Launch and configure the NFS filesystem as a service.

4. Launch and configure the load balancer service.

Now that we have the core resources in our cloud provider, we can perform the following tasks:

1. Gather information on the services and resources we have launched so far in our cloud provider.

2. Dynamically generate the script needed to bootstrap the admin virtual machine instance.

3. Dynamically generate the script needed to bootstrap the web virtual machine instance(s).

   Once we have the scripts, we can continue to launch our workload.

4. Launch the admin virtual machine instance, attaching the script we generated; this should then do the following, once it is executed when the instance first boots:

    A. Run an operating system update

    B. Download, install, and configure Apache, PHP, and the MySQL clients and NFS clients

    C. Configure the remote NFS share by creating the mount points, setting it to mount on boot, and also ensuring that the NFS share is mounted before progressing any further

    D. Download WordPress, bootstrap the database, and configure the site

    E. Start the webserver and make sure that all the services we have installed and configured are configured to start after a reboot

Now that the admin virtual machine instance has been launched, we hopefully have a working copy of WordPress stored on our NFS share, meaning we can progress with the remaining web virtual machine instances:

1. Launch the web virtual machine instances, attaching the script we generated; this should then do the following, once it is executed when the instance boots:

    A. Run an operating system update

    B. Download, install, and configure Apache, PHP, and the MySQL clients and NFS clients

    C. Configure the remote NFS share by creating the mount points, setting it to mount on boot, and ensuring that the NFS share is mounted before progressing any further

    D. Start the webserver and ensure all the services we have installed and configured will start after a reboot

    We should now have – if everything has gone as planned – a working WordPress installation across a small number of virtual machine instances, which leaves us with one last task.

2. Register all the virtual machine instances with the load balancer so they can start receiving traffic.

There are a few other cloud provider-specific tasks, which we will get into in the next few chapters as we get lower into the design and start to write our Infrastructure as Code; that sums up the tasks we will need to complete in the rough order they need to be executed.

Back in *Chapter 2, Ansible and Terraform beyond the Documentation,* one of the points we mentioned was that Terraform isn't really designed to be used to deploy and configure the software – it's not that straightforward for it to SSH into a virtual machine host to install and configure the software stack, so how are we going to do that?

Let's answer that now, by using a tool called **cloud-init**.

Introducing cloud-init

In *steps 6* and *7* of the tasks we listed in the previous section, we talk about generating a script – this will be a `cloud-init` script. This is a cloud and Linux operating system-agnostic tool used for bootstrapping instances as they boot.

It is supported on both Microsoft Azure and Amazon Web Services, and we will be using our Infrastructure as Code tools to populate a base template with the information gathered on the resources that have been launched, such as SQL and NFS endpoints, and then attach the output to the virtual machine instances when they are launched.

What follows is an example `cloud-init` script, which, when deployed with a virtual machine instance, will do the following tasks:

1. Update all of the packages which are already installed on the virtual machine to ensure we are fully patched.

2. Install NGINX.

3. Create a default NGINX site.

4. Create an example `index.html` file and place it in the root of the default NGINX site we configured in *step 2*.

5. Restart the NGINX service to pick up the new configuration.

To action these steps, the script looks like the following:

```
#cloud-config
package_upgrade: true
packages:
  - nginx
write_files:
  - owner: www-data:www-data
    path: /etc/nginx/sites-available/default
    content: |
      server {
            listen              80 default_server;
            root                /var/www/site;
            index               index.html;
            try_files $uri /index.html;
      }
  - owner: www-data:www-data
    path: /var/www/site/index.html
    content: |
      <!DOCTYPE html>
      <html>
          <head>
```

```
                <title>Example</title>
        </head>
        <body>
                <p>This is an example of a simple HTML page.</p>
        </body>
    </html>
runcmd:
  - service nginx restart
```

As you can see, it is relatively straightforward to read and follow along with what is happening; the one we will be using to deploy WordPress is a little more complicated as it is going to do a lot more than the example we have just given – but more on that in *Chapter 4, Deploying to Microsoft Azure*, and *Chapter 5, Deploying to Amazon Web Services*.

> **Information**
>
> Please note that while the example above used NGINX as the webserver, we will use Apache for the web server in our Wordpress deployment.

This means that we will still be able to use the programmatic parts of Terraform to configure our virtual machine instances without having to SSH into them. For Ansible, in *Chapter 6, Building upon the Foundations*, we will be taking a slightly different approach where we will use SSH to log in to our virtual machine to be able to make changes to the software stack and its configuration.

Now that we know the steps we are going to be taking to deploy our workload, let's visualize what it is going to look like.

Exploring the high-level architecture

Now that we know what we are going to be deploying, we should have a good idea of what the high-level architecture is going to look like. The following diagram shows a cloud-agnostic overview of how the resources we are going to be deploying across the next two chapters are going to hang together:

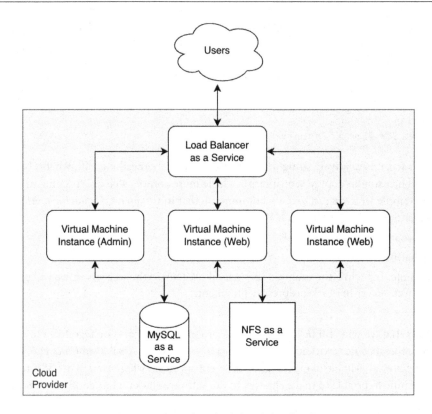

Figure 3.1 – An overview of our high-level cloud architecture

From a software stack perspective, each of the virtual machine instances will look like the following:

Figure 3.2 – An overview of our high-level software architecture

While this is not the most verbose high-level design, we now have a good idea of what it is we need to code to deploy our WordPress-based workload.

Summary

In this chapter, we have spoken about the example project we are going to look at deploying into Microsoft Azure and Amazon Web Services in the next two chapters using both Terraform and Ansible. While we have kept the discussion at a high level and aimed to be as cloud-agnostic as possible, we know the tasks we need to follow and roughly how we will get around the limitation of Terraform not really being a tool you can use to manage your application deployment.

Now that we know what our cloud and software architecture is going to look like, as well as having an idea of the order in which we need to deploy the resources, we can make a start on getting into the low-level design and the actual deployments. In our next chapter, we will look at deploying the workload discussed to Microsoft Azure.

Further reading

You can find more details on the software we have mentioned in this chapter at the following URLs:

- WordPress: `https://wordpress.org/`
- PHP: `https://www.php.net/`
- MySQL: `https://www.mysql.com/`
- NGINX: `https://nginx.org/`
- `cloud-init`: `https://cloud-init.io/`

Part 2:
Getting Hands-On
with the Deployment

Now that we have an understanding of the tools we are going to be using and also have an idea of the tasks we need to execute to deploy our example workload, it is time to roll up our sleeves and make a start on the code and deploy it.

In this part, we will be deploying the workload using Terraform and Ansible in Microsoft Azure and Amazon Web Services and discussing how we could build upon the scripts.

This part has the following chapters:

- *Chapter 4, Deploying to Microsoft Azure*
- *Chapter 5, Deploying to Amazon Web Services*
- *Chapter 6, Building upon the Foundations*

4

Deploying to Microsoft Azure

In this fourth chapter, we are going to look at getting our project deployed with the first of the two major public cloud providers we are going to cover in this book, **Microsoft Azure**.

We are going to cover the following topics:

- Introducing Microsoft Azure

- Preparing our cloud environment for deployment

- Producing the low-level design

- Terraform – writing the code and deploying our infrastructure

- Ansible – reviewing the code and deploying our infrastructure

We will delve into the world of Microsoft Azure, beginning with an introduction to the platform, its key features, and the benefits it offers for cloud-based application deployment. We will also explore the different services available within Azure and how they fit into our architectural design for our WordPress workload.

Following this, we will build on our Terraform knowledge and work through the code needed to provision and manage our Azure cloud infrastructure. Lastly, we will explore Ansible, another essential tool for automating infrastructure deployment and configuration management.

By the end of this chapter, you will have gained an understanding of Microsoft Azure and its various components and be equipped with the skills necessary to deploy and manage your applications on this cloud platform using Terraform and Ansible.

Technical requirement

Due to the amount of code needed to deploy our project, when it comes to the Terraform and Ansible sections of the chapter, we will not cover every piece of code needed to deploy the project. The code repository accompanying this title will contain the complete executable code.

Introducing and preparing our cloud environment

In 2008, Microsoft unveiled Windows Azure, a cloud-based data center service that had been in development under the internal project name *Project Red Dog*. This service included five core components:

- **Microsoft SQL Data Services**, a cloud version of the SQL database, which aimed to simplify hosting

- **Microsoft .NET Services**, a **Platform as a Service (PaaS)**, allowed developers to deploy their .NET-based applications in a Microsoft-managed runtime

- **Microsoft SharePoint and Dynamics**, **Software as a Service (SaaS)** versions of the company's intranet and customer relationship management products

- **Windows Azure** is an **Infrastructure-as-a-Service (IaaS)** offering that enables users to create virtual machines, storage, and networking services for their compute workloads

All the services provided by Microsoft as part of Windows Azure were built upon the Red Dog operating system, a specialized version of their Windows NT operating system, which had been specifically designed to include a cloud layer to support the delivery of data center services.

In 2014, the company decided to rebrand the service as Microsoft Azure; as they added services, it made sense for them to drop the Windows branding, especially as there was a growing number of Linux-based workloads being hosted on the platform. This trend continued over the following years, and by 2020, it was reported that more than half of Azure's virtual machine cores and a significant proportion of Azure Marketplace images were Linux-based, demonstrating Microsoft's increasing embrace of Linux and open source technologies as the building blocks for some of their now core services.

Now that we have some background knowledge of Microsoft Azure, let's start preparing the cloud environment for deployment.

Preparing our cloud environment for deployment

For the purposes of this chapter, we will run the Terraform and Ansible scripts locally on our own machine – this makes the deployment a little easier as we will be able to piggyback off a signed-in session using the Azure **Command-Line Interface (CLI)**. For details on how to install this, please see the official documentation at `https://learn.microsoft.com/en-us/cli/azure/install-azure-cli`.

Once installed, make sure you are signed into the account where you would like the resources to be deployed; you can do this by running the following command:

```
$ az login
```

Then, follow the on-screen prompts; if you are already logged in, then you can double-check the details of your current login by running this command:

```
$ az account show
```

Now that we have our environment prepared, we can now look at the services we are going to be deploying.

Producing the low-level design

Based on the deployment we discussed in *Chapter 3, Planning the Deployment*, we know we are going to need the following resources to run our workload on Microsoft Azure:

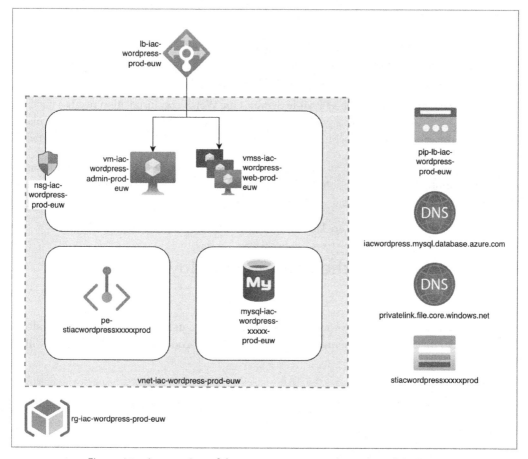

Figure 4.1 – An overview of the resources we are going to launch in Azure

We will use the following services:

- **Azure Load Balancer**: This is a TCP load balancer as a service – while I would have preferred to use **Azure Application Gateway** to terminate our HTTP/HTTPS connections, that would have added a little too much complexity to our build for this stage of the book.

- **Virtual Network**: The core networking service our services will be both deployed into or configured to be accessible from.

- **Virtual machine**: We will use a single **Linux virtual machine** as our WordPress admin instance – this will be responsible for the initial bootstrapping of the application.

- **Virtual Machine Scale Set**: This is similar to the Linux virtual machine, but this service is designed to manage one to many virtual machines from a single resource, allowing us to scale out if needed.

- **Azure Storage Account/Azure Files**: Our WordPress files will be stored in an NFS share, which is only accessible to trusted IP addresses within our Virtual Network where our virtual machine and Virtual Machine Scale Set instances are running.

- **Azure Database for MySQL - Flexible Server**: Our WordPress installation needs a database server; as we are running in a public cloud, a **Database-as-a-Service (DBaaS)** option makes sense. This service will make a MySQL server and database accessible within our Virtual Network.

There are also other services such as **Azure Private DNS**, **private endpoints**, **network security groups**, and **public IPs** within the solution to support securely accessing the core services we will launch within the Virtual Network.

Now that we know the services we are launching, we can dive into writing our code.

Terraform – writing the code and deploying our infrastructure

Now that we know which services we are going to deploy, we can make a start on our Terraform deployment. To make things more manageable, I will split our code into several files; they will be named the following:

- `001-setup.tf`

- `002-resource-group.tf`

- `003-networking.tf`

- `004-storage.tf`

- `005-database.tf`

- `006-vm-admin.tf`

- `007-vmss-web.tf`

- `098-outputs.tf`

- `099-variables.tf`

- `vm-cloud-init-admin.yml.tftpl`

- `vmss-cloud-init-web.tftpl`

I have done this to more logically group all the functions around a certain part of the deployment code together; for example, all of the networking elements can be found in the `003-networking. tf` file and the variables used in the `099-variables.tf` file.

> **Information**
>
> As mentioned at the start of this chapter, what follows is not 100% of the code contained within each of the files, and I will be referencing blocks out of the `variables` file in line with blocks from other files.

Without further delay, let's look at the Terraform code, starting with the setup tasks.

Setting up the Terraform environment

One of the first things we need to do is set up our Terraform environment for our deployment. To do this, we need to confirm which version of Terraform to use and which providers to download:

```
terraform {
  required_version = ">=1.0"
  required_providers {
    azurerm = {
      source  = "hashicorp/azurerm"
      version = "~>3.0"
    }
    azurecaf = {
      source = "aztfmod/azurecaf"
    }
    random = {
      source = "hashicorp/random"
    }
    http = {
      source = "hashicorp/http"
    }
  }
}
```

Now, we need to add a configuration block for one of the providers:

```
provider "azurerm" {
  features {}
}
```

While we are not putting any custom configuration in there, it must be present to progress with the deployment.

Finally, we come to the first task, which uses a module from the **Terraform Registry** to come up with all the Microsoft-defined variations of a region name, full name, short name, and so on.

To call the module and pass it is the region variable, use the following code:

```
module "azure_region" {
  source        = "claranet/regions/azurerm"
  azure_region  = var.location
}
```

The var.location variable we use is defined in the 099-variables.tf file as follows:

```
variable "location" {
  description = "Which region in Azure are we launching the resources"
  default     = "West Europe"
}
```

As you can see, we are setting default as West Europe; don't worry, if you don't want to launch your resources in that region, we will cover overriding the variables when executing the deployment in *Chapter 6, Building upon the Foundations*.

And that covers the 001-setup.tf file, now that we have all the basics in place, we can move on to creating the **resource group**.

Creating a resource group

As mentioned in earlier chapters, my day job sees me doing a lot of work in Microsoft Azure, and one of the things I adhere to is the **Cloud Adoption Framework**. This is a sensible documented set of recommendations around deploying resources into Microsoft Azure, which includes a naming scheme. Accordingly, one of the providers we are using helps dynamically create Azure resource names based on the information we pass it; we will use this throughout the deployment, as one of the provider's goals is to introduce naming consistency for nearly all of the resources we are going to deploy. The code to generate the name of the resource group looks like the following:

```
resource "azurecaf_name" "resource_group" {
  name          = var.name
  resource_type = "azurerm_resource_group"
  suffixes      = [var.environment_type, module.azure_region.location_
short]
  clean_input   = true
}
```

As you can see, we are passing several bits of information – three variables and dynamically generated bits of information, namely the following two variables:

```
variable "name" {
  description = "Base name for resources"
  default     = "iac-wordpress"
}
variable "environment_type" {
  description = "type of the environment we are building"
  default     = "prod"
}
```

And also, we are using the output Azure Region module, which will provide a short name of whichever region we define in our `variables` file at `099-variables.tf`. This is referenced as follows:

```
module.azure_region.location_short
```

The other important information we are passing is `resource_type`, which in our case, is `azurerm_resource_group`. This will give us an output that looks like the following figure, which we can then pass on to our next resource block:

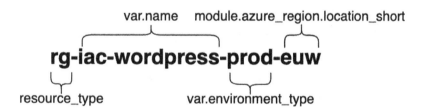

Figure 4.2 – Naming our resource group

Now that we have the resource name, we can go ahead and define the resource group block:

```
resource "azurerm_resource_group" "resource_group" {
  name     = azurecaf_name.resource_group.result
  location = module.azure_region.location_cli
  tags     = var.default_tags
}
```

As you can see, we reference `azurecaf_name.resource_group.result` as the name of the resource, and we also use another variation of the region name by using `module.azure_region.location_cli`, which will output the name as `westeurope` rather than `West Europe` or `euw`.

The final variable we pass in is for **resource tagging**. This variable is different from the ones we have passed so far as its type is map rather than string. This looks like the following:

```
variable "default_tags" {
  description = "The default tags to use across all of our resources"
  type        = map(any)
  default = {
    project     = "iac-wordpress"
    environment = "prod"
    deployed_by = "terraform"
  }
}
```

This is used throughout the deployment and will add three different tags, project, environment, and deployed_by, to each of the resources that use them. This is the simplest form of map we will use and is simply a list of keys and values.

Things will get a little more complicated with maps as we move on to the next section, which is *Networking*, as we start to use maps to introduce a little logic into our deployment.

Networking

As well as naming all of the resources using the azurecaf_name provider, we are going to configure and launch the following resources in this section:

- **Azure Virtual Network**, where we will configure the primary network resource, along with three subnets – which is where things start to get complicated with the maps

- **Azure Load Balancer**, as well as the resource itself, we will configure a public IP address, backend pool, health probe, and two types of rules – load balancing and NAT; there will be more on this later in the section

- **The network security group**, with two rules to allow secure access to our services

Let's dive into something more exciting and look at the Azure Virtual Network.

The Azure Virtual Network

The first part of configuring our underlying network is the Virtual Network resource itself. To do this, there are two main variables we are going to use; the first is straightforward:

```
variable "vnet_address_space" {
  description = "The address space of vnet"
  type        = list(any)
  default     = ["10.0.0.0/24"]
}
```

As can see, this defines the address space we are going to use for the Virtual Network as a list containing a single value; this is then called in the following block:

```
resource "azurerm_virtual_network" "vnet" {
  resource_group_name = azurerm_resource_group.resource_group.name
  location            = azurerm_resource_group.resource_group.location
  name                = azurecaf_name.vnet.result
  address_space       = var.vnet_address_space
  tags                = var.default_tags
}
```

Nothing too out of the ordinary on the face of it. To ensure that the Virtual Network is created after the resource group has been created, by passing in the dynamically generated name, the default tags, and the list of address spaces, which in our case only contains a single item, we use the following references:

- `azurerm_resource_group.resource_group.name`

- `azurerm_resource_group.resource_group.location`

The second variable is where we define the subnet and is also where things get interesting:

```
variable "vnet_subnets" {
  description = "The subnets to deploy in the vnet"
  type = map(object({
    subnet_name = string
    address_prefix = string
    private_endpoint_network_policies_enabled = bool
    service_endpoints = list(string)
    service_delegations  = map(map(list(string)))
  }))
```

The preceding code defines what variables are needed for each of the subnets, while the following code sets the default settings that we will use in our deployment, starting with the subnet for the virtual machines:

```
  default = {
    virtual_network_subnets_001 = {
      subnet_name = "vms"
      address_prefix = "10.0.0.0/27"
      private_endpoint_network_policies_enabled = true
      service_endpoints  = ["Microsoft.Storage"]
      service_delegations = {}
    },
```

The second subnet will be used for the private endpoints we will deploy:

```
    virtual_network_subnets_002 = {
      subnet_name  = "endpoints"
      address_prefix = "10.0.0.32/27"
      private_endpoint_network_policies_enabled = true
      service_endpoints = ["Microsoft.Storage"]
      service_delegations = {}
    },
```

The third and final subnet we are going to need is the one used for the `database` service:

```
    virtual_network_subnets_003 = {
      subnet_name = "database"
      address_prefix = "10.0.0.64/27"
      private_endpoint_network_policies_enabled = true
      service_endpoints = ["Microsoft.Storage"]
      service_delegations = {
        fs = {
          "Microsoft.DBforMySQL/flexibleServers" = ["Microsoft.
Network/virtualNetworks/subnets/join/action"]
        }
      }
    },
  }
}
```

As you can see, a lot is happening here, so let's break it down a little.

What we are defining here is a map that contains several objects; those objects are strings, a Boolean, a list, and finally, a map made up of a map, which contains a list of strings!

Let's start simple and look at how we will name the subnets. To do this, we use a `for_each` loop:

```
resource "azurecaf_name" "virtual_network_subnets" {
  for_each     = var.vnet_subnets
  name         = each.value.subnet_name
  resource_type = "azurerm_subnet"
  suffixes     = [var.name, var.environment_type, module.azure_
region.location_short]
  clean_input  = true
}
```

This takes the subnet_name value from each of our three maps and creates three resource names; the following names will be generated:

- snet-endpoints-iac-wordpress-prod-euw

- snet-vms-iac-wordpress-prod-euw

- snet-database-iac-wordpress-prod-euw

We will take a similar approach by using a for_each loop to create the subnets, but this time more of the information in the map object. The code block to create the subnets looks like the following:

```
resource "azurerm_subnet" "vnet_subnets" {
  for_each                = var.vnet_subnets
  name                    = azurecaf_name.virtual_network_subnets[each.
key].result
  resource_group_name     = azurerm_resource_group.resource_group.name
  virtual_network_name    = azurerm_virtual_network.vnet.name
  address_prefixes        = [each.value.address_prefix]
  service_endpoints = try(each.value.service_endpoints, [])
  private_endpoint_network_policies_enabled = try(each.value.private_
endpoint_network_policies_enabled, [])
  dynamic "delegation" {
    for_each = each.value.service_delegations
    content {
      name = delegation.key
      dynamic "service_delegation" {
        for_each = delegation.value
        iterator = item
        content {
          name = item.key
          actions = item.value
        }
      }
    }
  }
}
```

As already mentioned, we use the for_each argument to iterate over the elements in the vnet_subnets map. We set the name property using the results of the azurecaf_name loop; each of the three names is referenced using the key, which in the map would be as follows:

- virtual_network_subnets_001

- virtual_network_subnets_002

- virtual_network_subnets_003

Although, in this instance, we do not have to hardcode each of these values so we can use `azurecaf_name.virtual_network_subnets[each.key].result`. The `resource_group_name` property is set using the output of `azurerm_resource_group`. The `address_prefixes` property is set to a list containing the `address_prefix` value for the current subnet's `vnet_subnets` key.

The `service_endpoints` property is set to the corresponding value for the current subnet if it is provided. If the value is not provided, an empty list is used instead.

Similarly, the `private_endpoint_network_policies_enabled` property is set to the corresponding value for the current subnet if it is provided. If the value is not provided, an empty list is used instead. Finally, the code includes a nested loop over the `service_delegations` property of the current subnet.

A **dynamic block** is a setting block that is evaluated at runtime, allowing you to add blocks based on the values of input variables and also outputs from other tasks within the Terraform code. Dynamic blocks are helpful when you need to create a variable number of blocks based on data that is known only at runtime.

The dynamic block creates a `delegation` block for each element in the `service_delegations` map. The `iterator` argument is set to `item`, representing the current element being processed, and the `content` block creates a `service_delegation` block with the key as the name and the value as the actions.

To give you an idea of what this looks like, if we were to manually define the `virtual_network_subnets_002` and `virtual_network_subnets_003` map objects, they would look like the following:

```
resource "azurerm_subnet" "virtual_network_subnets_002" {
    name                    = "snet-endpoints-iac-wordpress-prod-euw"
    resource_group_name     = "rg-iac-wordpress-prod-euw"
    virtual_network_name = "vnet-iac-wordpress-prod-euw"
    address_prefixes        = ["10.0.0.32/27"]
    service_endpoints       = ["Microsoft.Storage"]
    private_endpoint_network_policies_enabled = true
}

resource "azurerm_subnet" "virtual_network_subnets_003" {
    name                    = "snet-database-iac-wordpress-prod-euw"
    resource_group_name     = "rg-iac-wordpress-prod-euw"
    virtual_network_name = "vnet-iac-wordpress-prod-euw"
    address_prefixes        = ["10.0.0.64/27"]
    service_endpoints       = ["Microsoft.Storage"]
    private_endpoint_network_policies_enabled = true
    delegation {
        name = "fs"
        service_delegation {
```

```
            name = "Microsoft.DBforMySQL/flexibleServers"
            actions = ["Microsoft.Network/virtualNetworks/subnets/join/
  action"]
        }
    }
}
```

As you can see, while it looks complicated, it is an excellent way of writing less code with hardcoded values, which makes the block easily reusable.

We will reference the output of the `for_each` loops when we add our network security group, and also when we start to attach resources to our subnets when we launch resources and place them in the subnets we have just defined.

The next lot of blocks launch and configure the Azure Load Balancer service; there is not much going on here that we haven't already covered, but here is a quick overview of each of the blocks:

- `"azurerm_public_ip"` `"load_balancer"` creates a public IP address

- `"azurerm_lb"` `"load_balancer"` launches the load balancer itself and attaches the public IP address we just launched

- `"azurerm_lb_backend_address_pool"` `"load_balancer"`, creates a backend pool; when we launch our virtual machine and Virtual Machine Scale Set instances, we will attach them to this backend pool

- `"azurerm_lb_probe"` `"http_load_balancer_probe"` adds a health probe to check that port 80 is open and accessible via a simple TCP test

- `"azurerm_lb_rule"` `"http_load_balancer_rule"` creates a rule to evenly distribute incoming requests on port 80, i.e., HTTP requests, across the instances in the backend pool if they are showing as healthy

- `"azurerm_lb_nat_rule"` `"sshAccess"` creates a rule that dynamically maps 2222 > 2232 to port 22 on the backend instances, giving us SSH to each of the instances in the backend pool

The final few tasks to do with the network are to create and configure a network security group; there is not much going on with the first few tasks:

- `"azurerm_network_security_group"` `"nsg"` creates the network security group

- `"azurerm_network_security_rule"` `"AllowHTTP"` adds a rule to the network security group we just created to allow HTTP access on port 80; this will be open to everyone

Next, we need to add a rule to allow SSH access to our hosts, but SSH is not a service that you would want to expose to the whole of the internet – even if we are going to access the instances using a non-standard port (remember we are using a NAT rule on Azure Load Balancer to map ports 2222 > 2232 to port 22 on the instances). So, we are going to use a data source to get the public IP address of the host that is currently running Terraform.

The block to get the current IP address uses the HTTP Terraform provider and looks like the following:

```
data "http" "current_ip" {
  url = "https://api.ipify.org?format=json"
}
```

As you can see, we are calling the `https://api.ipify.org/?format=json` URL, which returns a blob of JSON containing your current public IP address.

We can then take this blob and combine this with the `network_trusted_ips` variable, which is empty by default, containing a list of other trusted IP addresses:

```
variable "network_trusted_ips" {
  description = "Optional list if IP addresses which need access, your
current IP will be added automatically"
  type        = list(any)
  default = [
  ]
}
```

Now that we have the JSON containing our IP address and an optional list of other IP addresses we want to allow, we can create the rule itself:

```
resource "azurerm_network_security_rule" "AllowSSH" {
  name        = "AllowSSH"
  description = "Allow SSH"
  priority    = 150
  direction   = "Inbound"
  access      = "Allow"
  protocol    = "Tcp"
  source_address_prefixes     = setunion(var.network_trusted_ips,
["${jsondecode(data.http.current_ip.response_body).ip}"])
  source_port_range       = "*"
  destination_port_range  = "22"
  destination_address_prefix = "*"
  resource_group_name         = azurerm_resource_group.resource_group.
name
  network_security_group_name = azurerm_network_security_group.nsg.
name
}
```

As you can see, everything is happening in the `source_address_prefixes` entry; here, we use the built-in `setunion` function, which merges the content of `var.network_trusted_ips` – in our case, an empty list – and the JSON returned in the body of the request we made using the HTTP provider.

I have updated the relevant code slightly to make it a little easier to read:

```
setunion(
  var.network_trusted_ips,
  ["${jsondecode(data.http.current_ip.response_body).ip}"]
)
```

We are using `var.network_trusted_ips` as it is because that is already defined as a list; however, our IP address isn't, so we create a list using `[ ]` and then add an inline variable in Terraform; this is defined using `${ something here }`. `something here`, in our case, uses the built-in `jsondecode` function, which takes the body of the response that is held in `data.http.current_ip.response_body` and the value of the `ip` key, which is our public IP address.

Now that we have our network security group, we need to attach it to the subnet, which will host the virtual machine and Virtual Machine Scale Set instances. To do this, we need the ID of the subnet. To make this simple, I have created three variables with the name of the objects of each of the subnet maps:

```
variable "subnet_for_vms" {
  description = "Reference to put the virtual machines in"
  default     = "virtual_network_subnets_001"
}
variable "subnet_for_endpoints" {
  description = "Reference to put the private endpoints in"
  default     = "virtual_network_subnets_002"
}

variable "subnet_for_database" {
  description = "Reference to put the database in"
  default     = "virtual_network_subnets_003"
}
```

Now that we know the name of the object, which is the key name, we can use this in the block that associates the network security group with the subnet hosting the instances:

```
resource "azurerm_subnet_network_security_group_association" "nsg_
association" {
  subnet_id = azurerm_subnet.vnet_subnets["${var.subnet_for_vms}"].id
  network_security_group_id = azurerm_network_security_group.nsg.id
}
```

That concludes the network portion of the deployment, and we now have the underlying base to start deploying resources into, starting with storage.

Storage

We need to create a storage account with NFS enabled; most of these tasks are pretty straightforward:

- `"azurecaf_name"` `"sa"` generates the name of the storage account; there is a slight difference in that we are telling it to add a random string – we are doing this because storage account names must be unique across Azure, so if we didn't add this, the code might error depending on who else has already executed it

- `"azurecaf_name"` `"sa_endpoint"` takes the previous result and then generates a name for the private endpoint, which will be placed in the Virtual Network

- `"azurerm_storage_account"` `"sa"` creates the storage account itself

We now need to lock the storage account down so that only the three subnets in our Virtual Network and our trusted IP addresses have access – to do this, we use a block that looks like the following:

```
resource "azurerm_storage_account_network_rules" "sa" {
  storage_account_id = azurerm_storage_account.sa.id
  default_action     = var.sa_network_default_action
  ip_rules           = setunion(var.network_trusted_ips,
["${jsondecode(data.http.current_ip.response_body).ip}"])
  bypass             = var.sa_network_bypass
  virtual_network_subnet_ids = [
    for subnet_id in azurerm_subnet.vnet_subnets :
    subnet_id.id
  ]
}
```

As you can see, the IP rules use the same logic we employed when adding our public IP address to the network security group rule to allow us access to the instances using SSH access from a trusted IP address.

The part of the block to get the subnet IDs uses a `for` loop to iterate over the `azurerm_subnet.vnet_subnets` list, and for each subnet in the list, it extracts the `id` attribute of the subnet and adds it to the list of `virtual_network_subnet_ids`.

Once we have the network rules in place for the storage account, we can add the NFS share itself:

```
resource "azurerm_storage_share" "nfs_share" {
  name                 = replace(var.name, "-", "")
  storage_account_name = azurerm_storage_account.sa.name
  quota                = var.nfs_share_quota
  enabled_protocol     = var.nfs_enbled_protocol
```

```
  depends_on = [
    azurerm_storage_account_network_rules.sa
  ]
}
```

As you can see, we are use `depends_on` to ensure that the network rules are configured. We must declare `depends_on` as there is no output we can reference from `"azurerm_storage_account_network_rules"` `"sa"` when executing `"azurerm_storage_share"` `"nfs_share"`.

> **Hint**
>
> We will use small number of `depends_on` in the deployment. However, it is considered a best practice to keep its usage to a minimum and let Terraform figure out the dependencies as much as possible, as overuse of `depends_on` can slow down execution.

The keen-eyed among you may have also noticed that we are doing something different when naming this resource; we use the built-in `replace` function, taking the contents of the `var.name` variable and stripping out any hyphens.

The remaining tasks are all pretty basic:

- `"azurerm_private_dns_zone"` `"storage_share_private_zone"` creates a private DNS zone for `privatelink.file.core.windows.net`

- `"azurerm_private_dns_zone_virtual_network_link"` `"storage_share_private_zone"` takes the private DNS zone we created and attaches it to our Virtual Network

- `"azurerm_private_endpoint"` `"storage_share_endpoint"` creates the private endpoint, putting in the subnet defined in the `var.subnet_for_endpoints` variable and registering it with the private DNS zone

That concludes the storage; next up, we have the database service.

Database

We are now starting to get into the stride of things as we come to launch our database service:

- `"azurecaf_name"` `"mysql_flexible_server"` generates the name of the Azure MySQL Flexible Server

- `"azurecaf_name"` `"database"` generates the name of the database we will use for WordPress

- `"azurerm_private_dns_zone"` `"mysql_flexible_server"` adds the private DNS zone we will use for our Azure MySQL Flexible Server

- `"azurerm_private_dns_zone_virtual_network_link"` `"mysql_flexible_server"` registers the private DNS with our Virtual Network

Before we launch the **Azure MySQL Flexible Server**, there is one more thing we need to do, and that is to create a password. Rather than passing this using a variable, we can use the random provider to generate one based on the parameters we provide programmatically.

The following block will generate a random password 16 characters long and not use any special characters:

```
resource "random_password" "database_admin_password" {
  length  = 16
  special = false
}
```

There isn't too much out of the ordinary with the remaining blocks we use to launch and configure our Azure MySQL Flexible Server instance; what follows is a brief summary of what each of the blocks does:

- `"azurerm_mysql_flexible_server"` `"mysql_flexible_server"` launches the flexible server; we use `depends_on` here to ensure that the DNS zone is registered with the Virtual Network; otherwise, we could get an error when it came to creating the resource

- `"azurerm_mysql_flexible_server_configuration"` `"require_secure_transport"` changes the Azure MySQL Flexible Server parameter to allow non-TLS connections

- `"azurerm_mysql_flexible_database"` `"wordpress_database"` creates a database hosted on the Azure MySQL Flexible Server; once the parameter has been updated, we use `depends_on` to achieve this

Now that we have our Azure MySQL Flexible Server instance configured and connected to the Virtual Network, we are ready to launch our admin virtual machine instance.

The admin virtual machine

The admin virtual machine is going to be a single Linux virtual machine instance, which will be used to bootstrap our WordPress installation. Firstly, there are no new techniques used here, so rather than go into detail, here is an overview of what each of the blocks does:

- `"azurecaf_name"` `"admin_vm"` generates the name of the virtual machine

- `"azurecaf_name"` `"admin_vm_nic"` generates the name of the network interface

- `"azurerm_network_interface"` `"admin_vm"` creates the network interface resource, attaching it to the subnet defined in the `var.subnet_for_vms` variable

- `"random_password"` `"wordpress_admin_password"` generates a random password for the WordPress admin area – this time, using special characters apart from _%@

- `"random_password"` `"vm_admin_password"` generates the password for the virtual machine instances; this time, it's a little more complicated as virtual machines have password strength requirements, so we are going to generate a 16-character password with a minimum of two uppercase and two lowercase letters, two special characters, excluding !@#$%&, and also numbers

The next task, `"azurerm_linux_virtual_machine"` `"admin_vm"`, launches the virtual machine itself, and for the most part, there isn't much interesting going on with it apart from the section where we pass in `user_data`, which is where the `cloud-init` script is generated and passed over to Azure to inject when the virtual machine is launched. This part of the block looks like the following:

```
  user_data = base64encode(templatefile("vm-cloud-init-admin.yml.
tftpl", {
    tmpl_database_username = "${var.database_administrator_login}"
    tmpl_database_password = "${random_password.database_admin_
password.result}"
    tmpl_database_hostname = "${azurecaf_name.mysql_flexible_server.
result}.${replace(var.name, "-", "")}.mysql.database.azure.com"
    tmpl_database_name     = "${azurerm_mysql_flexible_database.
wordpress_database.name}"
    tmpl_file_share        = "${azurerm_storage_account.sa.name}.
file.core.windows.net:/${azurerm_storage_account.sa.name}/${azurerm_
storage_share.nfs_share.name}"
    tmpl_wordpress_url     = "http://${azurerm_public_ip.load_
balancer.ip_address}"
    tmpl_wp_title          = "${var.wp_title}"
    tmpl_wp_admin_user     = "${var.wp_admin_user}"
    tmpl_wp_admin_password = "${random_password.wordpress_admin_
password.result}"
    tmpl_wp_admin_email    = "${var.wp_admin_email}"
  }))
}
```

Let us dig into what is going on here a little bit more. First off, we need to pass the `cloud-init` script as Base64-encoded; luckily, Terraform has the `base64encode` function we can use to do this.

> **Information**
>
> Base64 is a way to encode data into a continuous string of ASCII text; it helps post-multi-line scripts or binary data to APIs. It is not a secure way to encode data since it can be easily decoded and does not provide any form of encryption. If we were to encode `Hello, world!`, it would be encoded as `SGVsbG8sIHdvcmxkIQ==` in Base64. `==` at the end is added to pad the string and make it a multiple of four characters since Base64 encoding works with blocks of four characters.

The next part uses Terraform's native `templatefile` function to read a file, which in our case, is called `vm-cloud-init-admin.yml.tftpl`. Once that has been defined, we pass a list of variables to use within the template and their values – here we are passing in details of the following:

- The Azure MySQL Flexible Server
- The Azure Files hosted NFS share
- The URL, which is made up of the public IP address of Azure Load Balancer
- Our WordPress installation information, which we have defined as variables in our main `variables` file

To save confusion, I am prefixing each of the variables we are passing into the template file with `tmpl`; this is not a requirement, but I find it helpful to distinguish between variables I can use in the main Terraform blocks and ones used within the templates.

An abridged version of the `cloud-init` template file is given in the following code block; it contains the bits that mount the NFS share:

```
#cloud-config
package_update: true
package_upgrade: true
packages:
  - nfs-common
runcmd:
  - sudo mount -t nfs ${tmpl_file_share} /var/www/html -o
vers=4,minorversion=1,sec=sys
  - echo "${tmpl_file_share} /var/www/html nfs
vers=4,minorversion=1,sec=sys" | sudo tee --append /etc/fstab
```

As you can see, the syntax for referencing the variables is slightly different from the main Terraform blocks, as we do not have to reference them as a variable with `${var.something}` and instead can just use `${something}`.

The fully rendered file is then passed to the virtual machine, and the script executes once the virtual machine boots. The full `cloud-init` file performs the following tasks:

- Updates all installed packages
- Installs the packages we need to run WordPress – for example, Apache2, PHP, and the NFS and MySQL client software
- Mounts the NFS share and adds a line to `/etc/fstab` so it mounts automatically after the instance reboots
- Installs the WordPress command-line client
- Sets up the correct permissions on the folders where we will be installing WordPress

- Downloads the latest version of WordPress
- Creates a `wp-config.php` file populated with the details of our Azure MySQL Flexible Server
- Installs WordPress itself using the variables we passed in
- Tweaks the Apache configuration and restarts the service

Once these steps have been completed, we should have a working WordPress installation.

A Web Virtual Machine Scale Set

Now that we have our admin virtual machine instance bootstrapped, we can launch a Virtual Machine Scale Set to act as web servers. As we already have a WordPress installation running and all the files needed to serve the website on the NFS share, these instances only need to have the basic software stack configured.

Also, as most of the heavy lifting has already been completed, this should be straightforward:

- `"azurecaf_name"` `"web_vmss"` generates the name of the Virtual Machine Scale Set.
- `"azurecaf_name"` `"web_vmss_nic"` generates the name of the network interface used by the virtual machine scale set.
- `"azurerm_linux_virtual_machine_scale_set"` `"web"` creates the virtual machine scale set itself; this is in line with the admin virtual machine we launched, and we are reusing many of the same variables, with the notable addition of `var.number_of_web_servers`, which defines the number of server instances to launch. We are also using a cutdown version of the `cloud-init` script called `vmss-cloud-init-web.tftpl`.

That concludes launching and configuring our Azure resource; there is just a small number of blocks left before we are finished.

Output

The final file outputs some helpful information about our deployment. As some of the output includes information, for example, the results of `random_password`, we will need to mark that part of the output as `sensitive`, as we don't want our randomly generated password to be visible when logging our output, which is being printed to the screen, for example:

```
output "wp_password" {
  value    = "Wordpress Admin Password: ${random_password.wordpress_
admin_password.result}"
  sensitive = true
}
output "wp_url" {
  value    = "Wordpress URL: http://${azurerm_public_ip.load_
balancer.ip_address}/"
```

```
    sensitive = false
}
```

Now that we have an understanding of what the Terraform code is going to do, we can now run it.

Deploying the environment

To deploy the environment, we simply need to run the following commands:

```
$ terraform init
$ terraform plan
```

When you run the `terraform plan` command, it will give you an overview of what resources are going to be deployed, as well as some very basic error-checking to make sure that everything is in order.

If you have the output of the plan, then you can proceed with the deployment by running the following command:

```
$ terraform apply
```

Upon completion, you should see something like the following screen:

```
● ● ●   russ@Russ-Laptop:~/Library/CloudStorage/OneDrive-Personal/Documents/PACKT/B19537_Infrastructure_as_Code_for_Beginners/Code/chapter04/terraform-azure    ⌥⌘1
azurerm_linux_virtual_machine_scale_set.web: Still creating... [10s elapsed]
azurerm_linux_virtual_machine_scale_set.web: Still creating... [20s elapsed]
azurerm_linux_virtual_machine_scale_set.web: Still creating... [30s elapsed]
azurerm_linux_virtual_machine_scale_set.web: Still creating... [40s elapsed]
azurerm_linux_virtual_machine_scale_set.web: Still creating... [50s elapsed]
azurerm_linux_virtual_machine_scale_set.web: Still creating... [1m0s elapsed]
azurerm_linux_virtual_machine_scale_set.web: Still creating... [1m10s elapsed]
azurerm_linux_virtual_machine_scale_set.web: Creation complete after 1m12s [id=/subscriptions/ce7aa0
b9-3545-4104-99dc-d4d082339a05/resourceGroups/rg-iac-wordpress-prod-euw/providers/Microsoft.Compute/
virtualMachineScaleSets/vmss-iac-wordpress-web-prod-euw]

Apply complete! Resources: 49 added, 0 changed, 0 destroyed.

Outputs:

vm_password = <sensitive>
wp_password = <sensitive>
wp_url = "Wordpress URL: http://20.23.249.255/"
 ~/Library/CloudStorage/OneDrive-Personal/Documents/PACKT/B19537_Infrastructure_as_Code_for_Beginner
s/Code/chapter04/terraform-azure ⟩ ⌥ main ⟩
```

Figure 4.3 – The finished Azure deployment

As you can see, the two password outputs have been marked as `sensitive`, but we have the URL of the WordPress installation. Now, let's dive straight into the WordPress admin portal, open the URL you have been given in your browser, and append `wp-admin` at the end of the URL. For example, for me, the URL was `http://20.23.249.255/wp-admin`.

> **Note**
> All URLs and passwords used in this chapter have long since been terminated and are not valid; please use the details from your own deployment.

This should give you a login page that looks like the following:

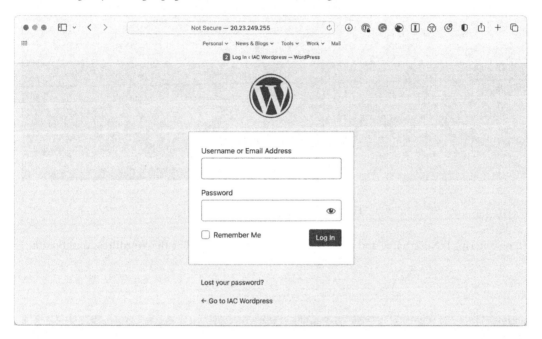

Figure 4.4 – The WordPress login screen

We know that the username for WordPress is admin, as we have that set as a variable, but how about that password? Well, by default, Terraform will always display sensitive when you run the terraform output command; however, appending -json at the end of that will give you the full output.

You can see the output of me running `terraform output -json` in the following screenshot:

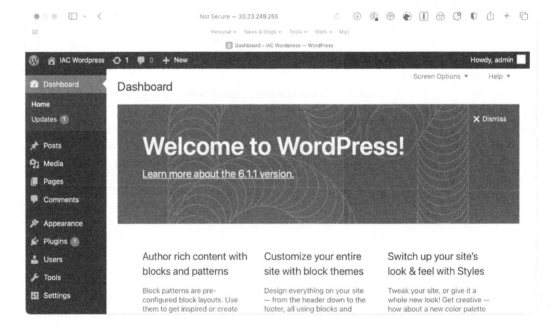

Figure 4.5 – Accessing the passwords

Upon entering the username and password, you should be greeted by the WordPress dashboard:

Figure 4.6 – The WordPress dashboard

You can also go to the Azure portal and check the resources there; you should be able to find them in the `rg-iac-wordpress-prod-euw` resource group (assuming that you have kept the variables at their defaults and have not updated them; if you have, then you will need to find the group that matches your updates).

Once you have finished, please don't forget to run the following command:

```
$ terraform destroy
```

Otherwise, you will incur costs for running resources if you have followed along and launched the environment in your own account.

Ansible – reviewing the code and deploying our infrastructure

While we have covered Terraform in detail in this chapter, we will only quickly review the Ansible code here as we will go into a lot more detail on an Ansible deployment in the next chapter, *Chapter 5, Deploying to Amazon Web Services*.

Like Terraform, the Ansible code is split into roles; this makes our `site.yml` file look like the following:

```
- name: Deploy and configure the Azure Environment
  hosts: localhost
  connection: local
  gather_facts: true
  vars_files:
    - group_vars/azure.yml
    - group_vars/common.yml
  roles:
    - roles/create-randoms
    - roles/azure-rg
    - roles/azure-virtualnetwork
    - roles/azure-storage
    - roles/azure-mysql
    - roles/azure-vm-admin
    - roles/azure-vmss-web
    - roles/output
```

Here, we load two variable files from the `group_vars` folder and calling eight different roles. As we have already discussed, Ansible will execute its tasks in the order we call them, so the ordering of the roles is essential.

Ansible Playbook roles overview

Let's dive straight in and look at the first role that is called in the `site.yml` file. The **randoms** role has a single function: randomly generating all the variables we will need for the deployment. However, this is where we hit our first significant difference between Ansible and Terraform. As Ansible is stateless, once it has generated a random value and the Ansible execution has stopped, it will instantly forget about it. This means that when we next execute our playbook, it will regenerate the random values, which, as we are using them in resource definitions, could cause a new resource to be launched.

What we need to do is create a file that contains all the random values, but if there is already a file there, then continue without updating them – the process we need to follow is visualized in the following workflow:

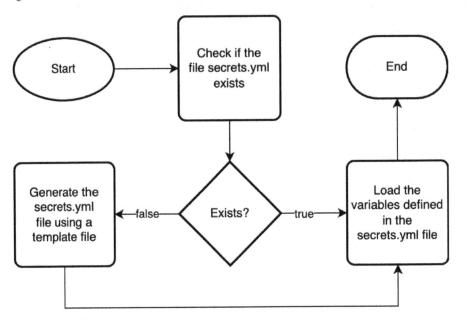

Figure 4.7 – Do we need to create a secrets.yml variable file?

So, what does this look like as Ansible tasks?

```
- name: Check if the file secrets.yml exists
  ansible.builtin.stat:
    path: "group_vars/secrets.yml"
  register: secrets_file
```

As you can see, we use the built-in `stat` module to check for the presence of a file in the `group_vars` called `secrets.yml`. We then register the results of this task as a variable called `secrets_file`.

The contents of the `secrets_file` variable we registered when the `secrets.yml` file doesn't exist on your filesystem, which it should do, as we are going to create the file shortly, look like the following:

```
TASK [roles/create-randoms : print the secrets_file variable] ********
*****************************
ok: [localhost] => {
    "msg": {
        "changed": false,
        "failed": false,
        "stat": {
            "exists": false
        }
    }
}
```

As you can see, there is a single output called `exists`, which is set to `false`.

So, our next task will generate the `secrets.yml` file if `exists` is set to `false` and looks like the following:

```
- name: Generate the secrets.yml file using a template file if not
exists
  ansible.builtin.template:
    src: "secrets.yml.j2"
    dest: "group_vars/secrets.yml"
  when: secrets_file.stat.exists == false
```

As you can see from the last line, the task will only run when `secrets_file.stat.exists` is equal to `false`; if it returns `true`, then the task will be skipped, as we do not need to generate the `secrets.yml` file again.

The task itself uses the `template` function to take the source template, which looks like the following, process it, and then output the rendered output to `group_vars/secrets.yml`:

```
short_random_hash: "{{ lookup('community.general.random_string',
length=5, upper=false, special=false, numbers=false) }}"
db_password: "{{ lookup('community.general.random_string', length=20,
upper=true, special=true, override_special="@-&*", min_special=2,
numbers=true) }}"
vm_password: "{{ lookup('community.general.random_string', length=20,
upper=true, special=true, override_special="@-&*", min_special=2,
numbers=true) }}"
wp_password: "{{ lookup('community.general.random_string', length=20,
upper=true, special=true, override_special="@-&*", min_special=2,
numbers=true) }}"
```

The template uses the `lookup` function to call the `random_string` module to generate a short random hash for resource names and various passwords.

> **Note**
>
> While this method works for our deployment, you should look at something more secure for a production environment – for example, dynamically loading the secret values from a remote key management store.

Now that we know the file is there, as we have just generated it, we can load the contents of the file in as variables:

```
- name: Load the variables defined in the secrets.yml file
  ansible.builtin.include_vars:
    file: "group_vars/secrets.yml"
```

Now that our variables are loaded, we can progress with the rest of the playbook run.

The resource group role

The **resource group role** has a single task, and that is to create the resource group – as you may have noticed, if you have already looked at the Ansible code in the GitHub repository for this chapter, there is no Ansible equivalent of the `azurecaf_name` provider we used in Terraform. Because of this, we have defined all the resource names in the `group_var/azure.yml` file using various variables and hardcoded values.

The Virtual Network role

There are several tasks in this role, a lot of which use some concepts we have yet to cover but will go into more detail in *Chapter 5, Deploying to Amazon Web Services*:

- `azure.azcollection.azure_rm_virtualnetwork` creates the Virtual Network
- `azure.azcollection.azure_rm_subnet` loops through and creates the subnets that don't have `database` in the name
- `azure.azcollection.azure_rm_subnet` loops through and creates the subnets that have `database` in the name
- `community.general.ipify_facts` gets the current IP address
- `ansible.builtin.set_fact` takes the preceding output and registers it as a fact
- `ansible.builtin.tempfile` creates an empty temporary file
- `ansible.builtin.template` dynamically generates the network security group tasks from a template file
- `ansible.builtin.include_tasks` loads and executes the network security group tasks we just generated

- `azure.azcollection.azure_rm_publicipaddress` creates a public IP address for use with the load balancer

- `azure.azcollection.azure_rm_loadbalancer` creates a load balancer

Now that we have all the underlying networking, we can launch the rest of the resources.

Storage roles

The tasks executed by this role are as follows:

- `azure.azcollection.azure_rm_resourcegroup_info` gets information on the resource group we have created

- `ansible.builtin.set_fact` uses some regular expressions to extract the subscription ID from the resource group ID and set a fact

- `ansible.builtin.tempfile` generates a temporary file, which will be used for the variables for the storage account rules

- `ansible.builtin.template` dynamically generates the variables containing the storage account network rules

- `ansible.builtin.include_vars` loads the variables we just generated

- `azure.azcollection.azure_rm_storageaccount` creates the storage account

- `azure.azcollection.azure_rm_resource` creates the NFS share

- `azure.azcollection.azure_rm_privatednszone` creates a private DNS zone

- `azure.azcollection.azure_rm_privatednszonelink` links the private DNS zone to our Virtual Network

- `azure.azcollection.azure_rm_subnet_info` gets information on the endpoint subnet

- `azure.azcollection.azure_rm_storageaccount_info` gets info on the storage account we just created

- `azure.azcollection.azure_rm_privateendpoint` creates a private endpoint using all of the information we have just gathered

- `azure.azcollection.azure_rm_privateendpointdnszonegroup` attaches the private endpoint to the private DNS zone

While that all seems straightforward, the keen-eyed among you may have noticed something which appears a little out of place.

When it came to creating the NFS share, the task we used was called `azure.azcollection.azure_rm_resource`, even though there is a module called `azure.azcollection.azure_rm_storageshare`; what gives?

At the time of writing, the `azure.azcollection.azure_rm_storageshare` module does not support the creation of NFS file shares on an Azure Storage account. So instead, we dynamically generate a payload and send it to the Azure Resource Manager REST API to create the resource. We will look at this in a little more detail in the next role.

The MySQL role

Azure Flexible Server for MySQL is another Azure service that doesn't have a native Ansible module, so we are going to have to use the REST API to not only create the server but also set the `require_secure_transport` parameter and create our WordPress database.

Before doing any of that, though, we need to create the DNS resources and gather a few bits of information on the networking resources we have already launched; the following tasks do so:

- `azure.azcollection.azure_rm_privatednszone` creates the private DNS zone for the database

- `azure.azcollection.azure_rm_privatednszonelink` attaches the DNS zone we just created to the Virtual Network

- `azure.azcollection.azure_rm_privatednszone_info` gets information on the private DNS we just created

- `azure.azcollection.azure_rm_subnet_info` gets information on the subnet we have created purely for use with the Azure Flexible Server for MySQL

We now have all the resources and information we need to create the Azure Flexible Server for MySQL. The task for doing this looks like the following:

```
- name: Create an Azure Flexible Server for MySQL using the REST API
  azure.azcollection.azure_rm_resource:
    api_version: "2021-05-01"
    resource_group: "{{ resource_group_name }}"
    provider: "DBforMySQL"
    resource_type: "flexibleServers"
    resource_name: "{{ database_server_name }}"
    body:
      location: "{{ location }}"
      properties:
        administratorLogin: "{{ database_config.admin_username }}"
        administratorLoginPassword: "{{ db_password }}"
        Sku:
          name: "{{ database_config.sku.name }}"
          tier: "{{ database_config.sku.tier }}"
        Network:
          delegatedSubnetResourceId: "{{ database_subnet_output.subnets[0].id }}"
```

```
        privateDnsZoneResourceId: "{{ database_private_dns_zone_
output.privatednszones[0].id }}"
      tags: "{{ common_tags }}"
```

The first part of the task, from `api_version` down to `body`, is used to construct the URL that we will call. The keys listed under `body` are the parameters and options that will be posted to the API endpoint URL, which we have dynamically created. The URL we will be posting to looks something like `https://management.azure.com/subscriptions/{subscriptionId}/resourceGroups/{resourceGroupName}/providers/Microsoft.DBforMySQL/flexibleServers/{serverName}?api-version=2021-05-01`.

Simple enough, you may think, but there is quite a large piece of logic that Ansible would typically take care of, which we now have to consider for ourselves.

As Ansible is just posting to the REST API, assuming the request is valid, it will get a `200` response back to say that the request was successful; it doesn't wait until the resource has been launched before `200` is returned, meaning that Ansible will happily move on to the next task, which will fail in our deployment because we are immediately making changes to the resource, which will have a state of `Creating`. To get around, this we have the following task:

```
- name: Wait for Azure Flexible Server for MySQL to be ready
  azure.azcollection.azure_rm_resource_info:
    api_version: "2021-05-01"
    resource_group: "{{ resource_group_name }}"
    provider: "DBforMySQL"
    resource_type: "flexibleServers"
    resource_name: "{{ database_server_name }}"
  register: database_wait_output
  delay: 15
  retries: 50
  until: database_wait_output.response[0] is defined and database_
wait_output.response[0].properties is defined and database_wait_
output.response[0].properties.state == "Ready"
```

This task will poll the REST API for information on our server every 15 seconds a maximum of 50 times until the REST API reports that the server `state` is `Ready`, and then it moves on to the next task:

- `azure.azcollection.azure_rm_resource` – now the server is ready, we can update the `require_secure_transport` parameter

- `azure.azcollection.azure_rm_resource` – finally, we can create the WordPress database

Now that we have launched and configured the database resources, we can launch some virtual machines – starting with the admin one.

The Admin Virtual Machine role

Compared to the Azure Flexible Server for MySQL role, this is quite straightforward and uses all Ansible-native Azure modules:

- `azure.azcollection.azure_rm_networkinterface` creates the network interface for use with the Virtual Machine
- `azure.azcollection.azure_rm_publicipaddress_info` gets information on the public IP address attached to the load balancer we launched – we need this for the `cloud-init` script
- `ansible.builtin.tempfile` creates a temporary file, which will store the `cloud-init` script
- `ansible.builtin.template` generates the `cloud-init` script for the admin server; like our Terraform deployment, this will install the packages and bootstrap WordPress
- `azure.azcollection.azure_rm_virtualmachine` launches the virtual machine using the resources created and configured previously

Now that we have our admin virtual machine, let's look at the web servers.

The Web Virtual Machine Scale Set role

This role contains three tasks:

- `ansible.builtin.tempfile` creates a temporary file, which will store the `cloud-init` script
- `ansible.builtin.template` generates the `cloud-init` script for the admin server, like our Terraform deployment; this will install the packages and not touch WordPress as it's already installed on our admin virtual machine
- `azure.azcollection.azure_rm_virtualmachinescaleset` creates the virtual machine scale set

The output role

This role simply outputs the details we need to access our WordPress installation; unlike Terraform, this information is visible when it runs.

This simple role, which just displays some text when running our playbook, is the final role, and we are now ready to run our playbook.

Running the Ansible Playbook

Now that we know what everything does, you can run the Ansible Playbook by running the following command:

```
$ ansible-playbook site.yml
```

Once finished, you should see something like the following output:

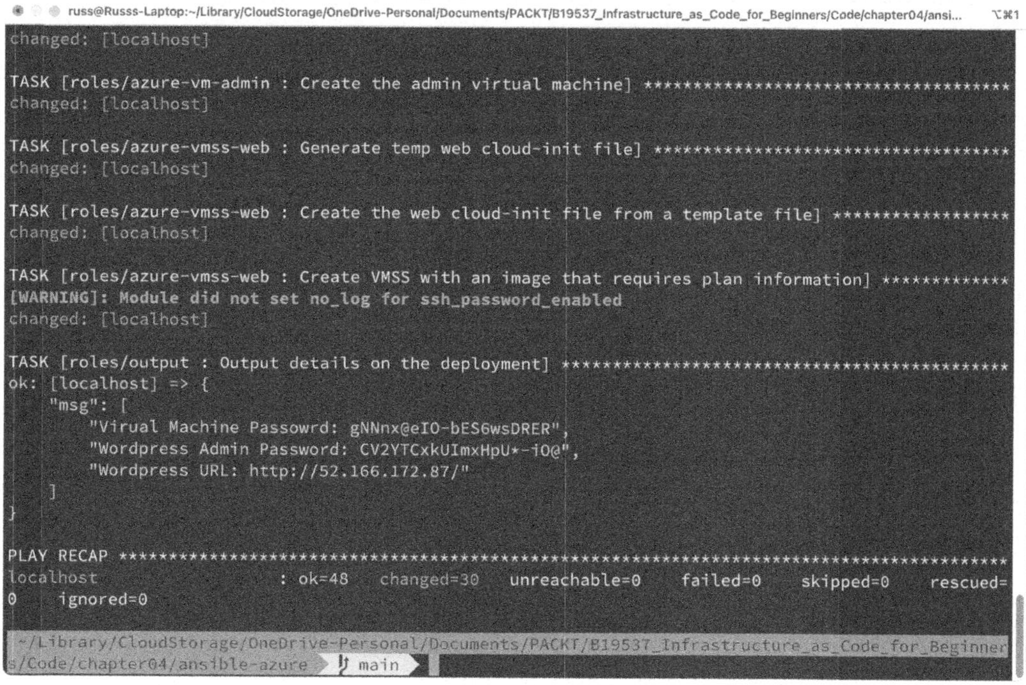

Figure 4.8 – The completed playbook run

Once you have looked at WordPress and the Azure resource, you can run the following command to remove the resource group and everything contained within it:

```
$ ansible-playbook destroy.yml
```

Remember to make sure that all of the resources have been removed by checking to see whether they are still listed in the Azure portal, as you could incur unexpected costs if the playbook failed for any reason.

Summary

In this chapter, we did a deep dive into using Terraform to deploy our WordPress environment in Microsoft Azure. We discussed the Terraform providers and worked through the Terraform code before finally executing it.

Also, as part of this walkthrough, we discussed some of the considerations you need to make when looping through resources, when it's appropriate to use depends_on, and how we can use templates to generate content.

Next up, we walked through the Ansible code, which deploys the same set of resources. This time, rather than a deep dive, we only went into detail on the Azure-specific details, as we will take a closer look at Ansible in *Chapter 5, Deploying to Amazon Web Services*.

Everything we covered so far should hopefully start to get you thinking about how you can apply some of the subjects we have covered to your own Infrastructure-as-Code deployments and you should already be starting to get a feel for which of the two tools you prefer.

In the next chapter, we will look at deploying our WordPress installation to Amazon Web Services, as well as doing a deeper dive into Ansible.

Further reading

You can find more details on the services and documentation we have mentioned in this chapter at the following URLs:

- Microsoft Azure: `https://azure.microsoft.com/`
- The Azure REST documentation: `https://learn.microsoft.com/en-us/rest/api/azure/`

Terraform providers and modules:

- azurerm: `https://registry.terraform.io/providers/hashicorp/azurerm/latest`
- azurecaf: `https://registry.terraform.io/providers/aztfmod/azurecaf/latest`
- random: `https://registry.terraform.io/providers/hashicorp/random/latest`

- HTTP: `https://registry.terraform.io/providers/hashicorp/http/latest`

- The Claranet Azure Region module: `https://registry.terraform.io/modules/claranet/regions/azurerm/latest`

A reference for Ansible collections is as follows:

- The Azure collection: `https://galaxy.ansible.com/azure/azcollection`

5

Deploying to Amazon Web Services

After deploying our WordPress infrastructure into Microsoft Azure, we are now ready to explore how to deploy the same infrastructure to **Amazon Web Services (AWS)**. However, while the high-level design of the infrastructure remains the same, there are some key differences between Azure and AWS that will require us to approach the deployment differently.

In *Chapter 4, Deploying to Microsoft Azure*, we focused on using Terraform to deploy to Azure. In this chapter, we will be diving deeper into Ansible, another popular infrastructure-as-code tool, to deploy our workload to AWS. Ansible allows us to define the desired state of our infrastructure in a declarative manner and manage the configuration and orchestration of our AWS resources.

By the end of this chapter, you will have a good understanding of how to use Ansible and Terraform to deploy a WordPress workload on AWS. You will also be familiar with the key differences between Azure and AWS and how to adapt your deployment approach to your infrastructure-as-code deployment.

We are going to be covering the following topics:

- Introducing Amazon Web Services
- Preparing our cloud environment for deployment
- Producing the low-level design
- Ansible – writing the code and deploying our infrastructure
- Terraform – reviewing the code and deploying our infrastructure

Technical requirements

Like in the previous chapter, due to the amount of code needed to deploy our project, when it comes to the Terraform and Ansible sections of the chapter, we will only be covering some pieces of code needed to deploy the project. The code repository accompanying this book will contain the complete executable code.

Introducing Amazon Web Services

AWS is a cloud infrastructure platform owned and operated by the e-commerce giant Amazon, which you probably already guessed, given the name.

The company began experimenting with cloud services in 2000, developing and deploying **application programming interfaces (APIs)** for their internal and external retail partners to consume. As more and more of the Amazon retail partners consumed more of the software services and grew at an exponential rate, they realized they would need to build a better and more standardized infrastructure platform to not only host the services they had been developing but also ensure that they could quickly scale as well.

Off the back of this requirement, Amazon engineers Chris Pinkham and Benjamin Black wrote a white paper, which Jeff Bezos personally approved in early 2004. The paper described an infrastructure platform where the compute and storage elements could all be deployed programmatically.

The first public acknowledgment of AWS's existence was made in late 2004. Still, at that time, the term was used to describe a collection of tools and APIs that would allow first and third parties to interact with Amazon's retail product catalog rather than the fully formed public cloud service it is today. It wasn't until 2006 that a rebranded AWS was launched, due mainly to the service starting to expand beyond offering an API to Amazon's retail services and instead starting to offer services that allowed users to use the services for their applications.

Simple Storage Service (S3) was the first of these new services; this was, and still is, albeit a little more feature-rich, a service that allows developers to write and serve individual files using a web API rather than having to write and read from a traditional local filesystem.

The next service to launch is also still around, Amazon **Simple Queue Service (SQS)**. It initially formed part of the original AWS collection of API endpoints. It is a distributed message system that again could be controlled and consumed by developers using an API.

The final service, launched in 2006, was a beta version of the Amazon **Elastic Compute Cloud (EC2)** service, which was limited to existing AWS customers – again, you could use the APIs developed by Amazon to launch and manage resources.

This was the final piece of the jigsaw for Amazon. They now had the foundations of a public cloud platform, which had initially been envisioned in the whitepaper that Chris Pinkham and Benjamin Black produced a few years earlier. They could not only use this new service for their retail platform but also sell space to other companies and the public, such as you and me. The bonus was that this new service could have a recurring revenue stream not only to pay for the initial development but also so that Amazon could maximize its hardware investment by *renting* out its idle compute resources.

AWS has grown from the 3 services mentioned in 2006 to over 200 in 2023. All these 200+ services are aligned with the core principles laid out in the original white paper. Each service is software-defined, meaning that a developer simply needs to make an API request to launch, configure, sometimes consume, and terminate the service.

Interacting with the API is also precisely what we will do in this chapter, as many of the principles laid out in that original white paper are also at the core of infrastructure as code.

Preparing our cloud environment for deployment

As we discussed in *Chapter 4*, *Deploying to Microsoft Azure*, we will be running Ansible and Terraform on our local machine, which means we can install and configure the AWS **command-line interface** (**CLI**).

Ansible and Terraform will use the credentials configured in the AWS CLI to authenticate against the AWS APIs. For details on how to install the AWS CLI, see `https://docs.aws.amazon.com/cli/latest/userguide/getting-started-install.html`.

Once installed, you need to generate and enter your credentials. This process is documented at `https://docs.aws.amazon.com/cli/latest/userguide/cli-configure-quickstart.html`.

Once configured, you should be able to run the following commands:

```
$ aws --version
$ aws ec2 describe-regions
```

When I run them on my own machine, I get the following output:

Figure 5.1 – The output of running the AWS version command to check the version

For the second command, there is quite a lot of output, which should look something like the following:

```
aws ec2 describe-regions                                        ⌥⌘1
{
    "Regions": [
        {
            "Endpoint": "ec2.ap-south-1.amazonaws.com",
            "RegionName": "ap-south-1",
            "OptInStatus": "opt-in-not-required"
        },
        {
            "Endpoint": "ec2.eu-north-1.amazonaws.com",
            "RegionName": "eu-north-1",
            "OptInStatus": "opt-in-not-required"
        },
        {
            "Endpoint": "ec2.eu-west-3.amazonaws.com",
```

Figure 5.2 – The output of running the AWS version command to describe the regions

Now that we have the AWS CLI configured and hooked into our AWS account, we can discuss the services we will be deploying and configuring within AWS.

Producing the low-level design

From an architecture point of view, the services that are going to be deployed are not too dissimilar from those we covered in *Chapter 4*, *Deploying to Microsoft Azure*:

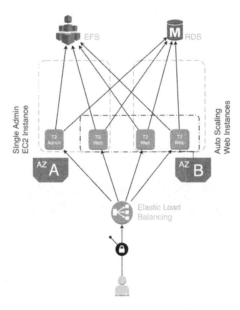

Figure 5.3 – An overview of the services we will be deploying into AWS

The core services we are going to be deploying are as follows:

- Amazon **Elastic Load Balancing** (**ELB**) is the first difference in the services we will be deploying. Azure Load Balancer only distributed TCP requests between our WordPress instances. However, in AWS, we will launch ELB configured as Application Load Balancer, which will terminate our HTTP requests and distribute them across our WordPress instances.

- **Amazon EC2** is the compute service. For our WordPress deployment, we will be deploying a single Amazon EC2 instance, which will be used to bootstrap WordPress, and then the rest of the Amazon EC2 instances will be auto-scaling.

- We will be using a combination of Amazon EC2 **Auto Scaling groups** (**ASGs**) and launch configurations to manage the deployment of the Amazon EC2 instances, which will host the web instances.

- Amazon **Elastic File System** (**EFS**) is the service that will provide the NFS share hosting of the WordPress installation, which will be shared across all of our instances.

- Amazon **Relational Database Service** (**RDS**) will be used to host the MySQL database we will use for WordPress.

- Amazon **Virtual Private Cloud** (**VPC**) is the underlying network service that hosts the services that will be deployed throughout this chapter. There are a few different services that come under this umbrella, and those will be covered in more detail when we do a deep dive into the Ansible code in the next section.

Now that we have an idea of the services we are going to be using and also which part of the WordPress deployment they will be hosting, we can start to look at how we approach the project using Ansible – this time going in into a little more detail than we did in *Chapter 4, Deploying to Microsoft Azure*.

Ansible – writing the code and deploying our infrastructure

In *Chapter 4, Deploying to Microsoft Azure*, we briefly covered the Ansible code to deploy our Azure environment. Let's take a step back and cover some of the basics we skipped.

While we can have one big YAML file containing our playbook code, I tend to split mine into more manageable chunks using roles. Roles can be used for a few things. In some cases, they can be stand-alone distributable tasks that can be reused across multiple projects; or, in our case, they can be used to manage more complex playbooks.

A copy of the `site.yml` file that will be used to deploy our WordPress environment in AWS looks like the following:

```
- name: Deploy and configure the AWS Environment
  hosts: localhost
```

```
connection: local
gather_facts: true
vars_files:
  - group_vars/aws.yml
  - group_vars/common.yml
roles:
  - roles/create-randoms
  - roles/aws-network
  - roles/aws-storage
  - roles/aws-database
  - roles/aws-vm-admin
  - roles/aws-asg-web
  - roles/output
```

As you can see, there are several roles, all of which I have grouped in the logical order we need to deploy our resources.

If you look in one of the role folders, for example, `roles/create-randoms`, you will notice that there are several folders and files:

- `defaults`: This is where the default variables for the role are stored. These can be overridden by any variables with the same name called in the `vars` folder.

- `files`: This folder contains any static files we wish to copy to the target hosts using the `copy` module.

- `handlers`: These are tasks that are executed once a playbook has been executed, for example, restarting services on a target host when a configuration file has changed.

- `meta`: This folder contains information about the role itself. This information would be used if it was ever published to Ansible Galaxy.

- `tasks`: This contains the primary set of instructions or actions that will be executed on the target hosts. These instructions are usually defined in YAML files, including installing packages, creating users, and copying files. Tasks can be organized into different files based on their functionality or the specific actions they perform. They can also include variables and conditional statements to make them more dynamic and flexible.

- `templates`: This folder contains the Jinja2 templates used by the `template` module.

- `tests`: If you are publishing your role to Ansible Galaxy, then it is a good idea to set up some tests. These are stored here.

- `vars`: You can override any of the variables defined in the `default` folder using the variables defined here. Variables defined here can also be overridden by any variables loaded from the `group_vars` folder at the top level of the playbook. These, in turn, can be overridden by variables passed in at runtime using the `ansible-playbook` command.

- README.md: This is the file used to create any documentation about the role when the role is checked into a service such as GitHub. This is useful when publishing the role to Ansible Galaxy.

Now that is a lot of folders and files to create when you want to add a role. Luckily, the `ansible-galaxy` command can bootstrap a role quickly. To do this, simply run the following command in the top level of your `playbook` folder, making sure to replace `role-name` with the name you want for your role:

```
$ ansible-galaxy init roles/role-name
```

This will create the folder and file structure we just covered and is an excellent starting point.

Before we dive into the roles, let's quickly discuss the variables. At the top of the `group_vars/aws.yml` file, we have some basic variables defined. These are as follows:

```
app:
  name: "iac-wordpress"
  location: "us-east-1"
  env: "prod"

wordpress:
  wp_title: "IAC WordPress"
  wp_admin_user: "admin"
  wp_admin_email: "test@test.com"
```

As you can see, we are defining a top level and have multiple keys and values attached to them. So, anywhere in our code, or even in other top-level variables, we can simply use something such as {{ app.name }}, which will be replaced by `iac-wordpress` when our playbook runs.

This can be seen when defining the resource names, as these are mostly made up of groups of variables defined elsewhere. Take the following example:

```
vpc_name: "{{ app.name }}-{{ app.env }}-{{ dict.vpc }}"
vpc_subnet_web01_name: "{{ app.name }}-{{ app.env }}-web01-{{ dict.
subnet }}"
vpc_subnet_web02_name: "{{ app.name }}-{{ app.env }}-web02-{{ dict.
subnet }}"
```

Now, let's look at the playbook roles in detail.

Ansible playbook roles

Let's dive straight in and look at the first role.

Creating the Randoms role

This role, which we already covered in detail in *Chapter 4, Deploying to Microsoft Azure*, does the same tasks as covered in that chapter. The role was copied straight from the Microsoft Azure deployment folder.

The AWS Network role

The primary variable we are defining in `group_vars/aws.yml` for this role is a lot more simplistic than the one we defined for the Azure deployment. It contains the CIDR range we want to use for our VPC network and nothing else:

```
vpc:
  address_space: "10.0.0.0/24"
```

The tasks we are running in the role take care of the rest of the information using some of Ansible's built-in functions. The first task is a relatively straightforward one:

```
- name: Create VPC
  amazon.aws.ec2_vpc_net:
    name: "{{ vpc_name }}"
    region: "{{ region }}"
    cidr_block: "{{ vpc.address_space }}"
    dns_hostnames: true
    dns_support: true
    state: present
  register: vpc
```

As you can see, it uses the `amazon.aws.ec2_vpc_net` module from the Amazon collection on Ansible Galaxy to create the VPC – so nothing too special or complicated there. The output of the task is registered as `vpc`; we will use this output register throughout the remainder of the playbook run.

The next task gathers some information on the region in which we are going to be deploying our workload:

```
- name: get some information on the available zones
  amazon.aws.aws_az_info:
    region: "{{ region }}"
  register: zones
```

Now that we have a few outputs registered, we can add the subnets and start doing more exciting things.

As part of our deployment, we need to add four subnets – two for the web services and two for the database services. Like our Azure deployment, the subnets will be /27s, and we will deploy each subnet in different Availability Zones.

> **Information**
>
> When we deployed the Azure version of our WordPress workload, we didn't have to worry about how the subnets were distributed across Availability Zones (which are different data centers within a region), as virtual networks in Azure can span multiple Availability Zones. However, AWS is different; subnets need to be pinned to Availability Zones, meaning you will need to have more than one per server role or service function.

The task to add the first subnet looks like the following:

```
- name: Create Subnet Web01
  amazon.aws.ec2_vpc_subnet:
    vpc_id: "{{ vpc.vpc.id }}"
    cidr: "{{ vpc.vpc.cidr_block | ansible.utils.ipsubnet(27, 0) }}"
    az: "{{ zones.availability_zones[0].zone_name }}"
    region: "{{ region }}"
    tags:
      Name: "{{ vpc_subnet_web01_name }}"
      Description: "{{ dict.ansible_warning }}"
      Project: "{{ app.name }}"
      Environment: "{{ app.env }}"
      Deployed_by: "Ansible"
  register: subnet_web01
```

Things start simple enough in that we use the output register from when we created the VPC to get the ID of the VPC to attach the subnet to by using `"{{ vpc.vpc.id }}"`.

Next, we use the output register again to get the CIDR range from the VPC output register; however, we take that value and use the `ansible.utils.ipsubnet` function to work out what the first `/27` in the CIDR range will be.

As we have passed in `10.0.0.0/24`, running `ansible.utils.ipsubnet(27, 0)` should give us `10.0.0.0/27`. The keen-eyed among you may have noticed that we passed in 0 rather than 1. Ansible always counts from 0, so if we had used 1, then we would have gotten `10.0.0.32/27`, which is what we need to use for the second subnet.

The second exciting thing we are doing is taking the output of the `zone` register, which contains information on the region we are using, including a list of the Availability Zones. So, when we use `{{ zones.availability_zones[0].zone_name }}`, it is taking the zone name for the first result, that is, 0.

The advantage of this approach in populating information for the CIDR and Availability Zone for the subnet is that we do not have to hardcode those details as variables. If we change the CIDR range or regions, the information would be programmatically generated to consider those changes.

When you write your Ansible playbooks, anything you can do to have your playbook adapt to user input or change dynamically is considered a best practice as it not only simplifies the information your consumers need to know but also makes the code reusable.

The rest of the task is populated using mostly static variables, so less interesting than what we have just covered.

This is then repeated for the second web subnet and the two subnets used with Amazon RDS – all we do is increment the numbers being passed to `ansible.utils.ipsubnet` and `zones.availability_zones`.

Once the subnets have been defined, we create an internet gateway using `amazon.aws.ec2_vpc_igw`. Following that, we create a route table to take advantage of the gateway using `amazon.aws.ec2_vpc_route_table`.

This is attached to the two web subnets and forwards all outgoing traffic to our internet gateway.

The next batch of tasks creates three security groups using the `amazon.aws.ec2_security_group` module.

The first of the three security groups will be assigned to the admin/web EC2 instances and the elastic load balancer. It opens ports `80` and `22` to the world, making them publicly accessible.

> **Information**
>
> For ease of use, I am opening port `22` to the world. In your production deployments, you should not do this and lock access down to one or more trusted IP addresses.

The next two security groups will be attached to the Amazon RDS and EFS services.

However, rather than defining source IP range(s), we are passing in the ID of the first security group we created, which means that ports `3306` (MySQL) and `2049` (NFS) will only be access to resources, which in our case is going to be the admin and web EC2 instances. The first security group attached will be able to access those services.

The final two tasks configure and launch the application load balancer. The first of the following two tasks is shown here and creates an empty ELB target group:

```
- name: Create an ELB target group
  community.aws.elb_target_group:
    name: "{{ alb_target_group_name }}"
    protocol: "HTTP"
    port: "80"
    vpc_id: "{{ vpc.vpc.id }}"
    region: "{{ region }}"
    state: "present"
    modify_targets: false
```

```
    tags:
      Name: "{{ alb_target_group_name }}"
      Description: "{{ dict.ansible_warning }}"
      Project: "{{ app.name }}"
      Environment: "{{ app.env }}"
      Deployed_by: "Ansible"
  register: alb_target_group
```

On the face of it, there does not appear to be anything too special about that, so why have I called it out?

At the time of writing, there is no module for creating an ELB target group using the amazon.aws collection; so instead, we have switched to using the community.aws collection. The developers use this collection as a staging ground for new features, and we will bounce between these two collections throughout the playbook.

Information

As modules may be promoted from being hosted in the community.aws collection to the amazon.aws collection in the future, please refer to the code in the GitHub repository that accompanies this book for the most recent updates.

The final task of the role is to create the application load balancer. Here, we use the amazon.aws. elb_application_lb module and several output registers we have created so far in the playbook run.

That concludes all the tasks we need to run to deploy the underlying network and supporting services. Now that we have those resources in place, we can deploy the storage for our WordPress installation.

The AWS Storage role

This is a simple role that contains a single task:

```
- name: Create the EFS resource
  community.aws.efs:
    name: "{{ efs_name }}"
    state: present
    region: "{{ region }}"
    targets:
      - subnet_id: "{{ subnet_web01.subnet.id }}"
        security_groups: ["{{ security_group_efs.group_id }}"]
      - subnet_id: "{{ subnet_web02.subnet.id }}"
        security_groups: ["{{ security_group_efs.group_id }}"]
    tags:
      Name: "{{ efs_name }}"
      Description: "{{ dict.ansible_warning }}"
      Project: "{{ app.name }}"
      Environment: "{{ app.env }}"
```

```
        Deployed_by: "Ansible"
    register: efs
```

As you can see, it uses the `community.aws.efs` module to create the Amazon EFS share, creating a target endpoint in our two web subnets. This step is important as EFS has a different DNS endpoint in each availability zone, so without this, we would be unable to connect to the NFS share in both of our web subnets.

The AWS Database role

Before we launch our EC2 instances, we need to have the MySQL Amazon RDS instance ready. This role contains two tasks – the first of which creates an RDS subnet group using the `amazon.aws.rds_subnet_group` module. Once we have the subnet group, the `amazon.aws.rds_instance` module is then used to create the RDS instance itself.

There isn't much to this role, but we now have an Amazon RDS instance and can start deploying our EC2 instances.

The AWS VM Admin role

Like in *Chapter 4, Deploying to Microsoft Azure*, we will deploy a single instance using a `cloud-init` script to bootstrap our WordPress installation.

The variables that we are going to be using for both this role and the next, which configures the ASG-managed EC2 instances, are as follows:

```
ec2:
  instance_type: "t2.micro"
  public_ip: true
  asg:
    min_size: 1
    max_size: 3
    desired_capacity: 2
  ami:
    owners: "099720109477"
    filters:
      name: "ubuntu/images/hvm-ssd/ubuntu-focal-20.04-amd64-server-*"
      virtualization_type: "hvm"
```

The first task that we will perform is generating a temporary file:

```
- name: Generate temp admin cloud-init file
  ansible.builtin.tempfile:
  register: tmp_file_create_cloud_init_admin_task
```

We will render the template file in `templates/vm-cloud-init-admin.yml.j2` and place the rendered contents in the temporary file we just created:

```
- name: Create the admin cloud-init file from a template file
  ansible.builtin.template:
    src: "vm-cloud-init-admin.yml.j2"
    dest: "{{ tmp_file_create_cloud_init_admin_task.path }}"
```

With the `cloud-init` file prepared, we can move on to the next step: figure out the ID of the **Amazon Machine Image** (**AMI**) we need to use:

```
- name: gather information about AMIs with the specified filters
  amazon.aws.ec2_ami_info:
    region: "{{ region }}"
    owners: "{{ ec2.ami.owners }}"
    filters:
      name: "{{ ec2.ami.filters.name }}"
      virtualization-type: "{{ ec2.ami.filters.virtualization_type }}"
  register: ubuntu_ami_info
```

As the AMI's maintainer, **Canonical**, which also develops Ubuntu, keeps the AMI up to date with patches and so on, a long list of AMIs will be returned as there are multiple versions.

Our next task sorts that list and takes the last item:

```
- name: filter the list of AMIs to find the latest one
  set_fact:
    ami: "{{ ubuntu_ami_info.images | sort(attribute='creation_date')
| last }}"
```

As you can see in the preceding code snippet, we are using the `sort` function to sort the list, which is JSON, by the `creation_date` attribute and then taking the `last` result. This leaves us with the details of the most recent AMI that Canonical has published.

Now we have everything we need to launch our admin EC2 instance:

```
- name: create the admin ec2 instance
  amazon.aws.ec2_instance:
    name: "{{ ec2_instance_name_admin }}"
    region: "{{ region }}"
    vpc_subnet_id: "{{ subnet_web01.subnet.id }}"
    instance_type: "{{ ec2.instance_type }}"
    security_group: "{{ security_group_web.group_name }}"
    network:
      assign_public_ip: "{{ ec2.public_ip }}"
    image_id: "{{ ami.image_id }}"
    user_data: "{{ lookup('file', tmp_file_create_cloud_init_admin_
```

```
task.path) }}"
    tags:
      Name: "{{ ec2_instance_name_admin }}"
      Description: "{{ dict.ansible_warning }}"
      Project: "{{ app.name }}"
      Environment: "{{ app.env }}"
      Deployed_by: "Ansible"
  register: ec2_instance_admin
```

As you can see, most all the information needed to deploy the instance is a variable, either a hardcoded one, such as `instance_type`, or one that is an output variable, like the one for `image_id`, which is the one we just gathered the information for.

For `user_data`, we are using the `lookup` function to read the contents of the temporary file we populated with the `cloud-init` script, which we will talk about in a moment.

Now that we have an EC2 instance, we need to register it in the ELB target group we created in the Network role, but we can only do this if the instance has a state of `running`.

Our Ansible playbook can progress a little too quickly, and the instance may not yet have entered that state, so we need to create a bit of logic that will pause the playbook execution and wait for the instance to have the correct state before progressing.

We can do this with the following task:

```
- name: Get information about the admin EC2 instance to see if its
running
  amazon.aws.ec2_instance_info:
    region: "{{ region }}"
    filters:
      instance-id: "{{ ec2_instance_admin.instances[0].instance_id }}"
  register: admin_ec2_instance_state
  delay: 5
  retries: 50
  until: admin_ec2_instance_state.instances[0].state.name == "running"
```

The task itself is quite simple; it uses the `amazon.aws.ec2_instance_info` module to gather information on the EC2 instance we have just launched.

On its own, this task would be pretty useless as it would gather the information once and then move on. The three lines at the end are bits that add the logic we need.

Using the `until` function, we take our output register, `admin_ec2_instance_state`, and check whether `state.name` is equal to `running`:

```
until: admin_ec2_instance_state.instances[0].state.name == "running"
```

If the `state.name` variable does not equal `running`, then retry 50 times every 5 seconds:

```
retries: 50
delay: 5
```

Continue until `state.name` equals `running`.

Once this condition is met, we know it will be safe to move on to the next task, and we won't get any errors because the state of the instance is incorrect:

```
- name: Update the ELB target group
  community.aws.elb_target_group:
    name: "{{ alb_target_group_name }}"
    protocol: "HTTP"
    port: "80"
    vpc_id: "{{ vpc.vpc.id }}"
    region: "{{ region }}"
    state: "present"
    modify_targets: true
    targets:
      - Id: "{{ ec2_instance_admin.instances[0].instance_id }}"
        Port: 80
```

So, we now have our admin EC2 instance running and registered with the ELB target group and the `cloud-init` script running. Well, sort of – we need to make a few adjustments to the `cloud-init` script from when we last looked at it in *Chapter 4, Deploying to Microsoft Azure*. Most of it is as we used it when deploying to Azure, with one additional piece of logic we needed to build in:

```
# Mount the NFS share and add it to fstab
  - until nc -vzw 2 {{ efs.efs.filesystem_address | regex_
replace("[^A-Za-z0-9.-]", "") }} 2049; do sleep 2; done; mount
-t nfs4 {{ efs.efs.filesystem_address }} /var/www/html -o
vers=4,minorversion=1,sec=sys
  - echo "{{ efs.efs.filesystem_address }} /var/www/html nfs4
vers=4,minorversion=1,sec=sys" | sudo tee --append /etc/fstab
```

As you can see from the preceding code snippet, there is a change to the line that mounts the NFS share provided by the Amazon EFS service – why have we needed to make this change?

If you remember, back when we launched the Amazon EFS service, we talked about the unique DNS endpoints being registered automatically in each subnet. To get around having to build the logic into our deployment to figure out which availability zone we are running our instances in so we can use the right DNS name for our Amazon EFS endpoint, a generic endpoint, CNAME, is created, which resolves to the appropriate endpoint for the subnet.

Great, you might be thinking that that saves us the hassle of having to code something to take this into account – and you would be correct, but it can take a while for this DNS alias to propagate.

As the `cloud-init` script is running completely independently of our Ansible playbook run, we can't use a conditional like the one that we just discussed for waiting, for instance, to have the correct state before progressing.

So, to get around this, we are adding the following:

```
until nc -vzw 2 somedns.domain.com 2049; do sleep 2;
done;
```

This is the Bash equivalent of the condition we added before. It will run the netcat (`nc`) command to see whether `somedns.domain.com` is responding on port `2049`. If it isn't, it will wait for two seconds using the `sleep` command and then repeat until we get the correct response.

You may have also noticed that we are using another Ansible function to get the details on the Amazon EFS endpoint from our output register.

By default, if we were to just use `{{ efs.efs.filesystem_address }}`, it would return the fully qualified domain name for our Amazon EFS endpoint with the filesystem path appended to the end of it, which in our case is `:/`.

This is not a valid address for the `nc` command to use, so we need to remove the `:/` from the address. To do this, we can use Ansible's `regex_replace` function, as we want to remove everything that isn't a regular character, dot, or hyphen. Then, this looks like the following:

```
{{ efs.efs.filesystem_address | regex_replace("[^A-Za-z0-9.-]", "") }}
```

This code should leave us with, for example, `somedns.domain.com` rather than `somedns.domain.com:/`.

The rest of the script remains intact. We also must use the same logic as previously for the cut-down version of the `cloud-init` script being used for the web EC2 instances being deployed using the ASG, which we will look at next.

The AWS ASG role

This role follows a similar pattern to the AWS VM Admin role, starting with generating the `cloud-init` script:

```
- name: Generate temp web cloud-init file
  ansible.builtin.tempfile:
  register: tmp_file_create_cloud_init_web_task

- name: Create the web cloud-init file from a template file
  ansible.builtin.template:
    src: "vm-cloud-init-web.yml.j2"
    dest: "{{ tmp_file_create_cloud_init_web_task.path }}"
```

We then need to create a launch configuration:

```
- name: Create launch config
  community.aws.autoscaling_launch_config:
    name: "{{ lauch_configuration_name }}"
    image_id: "{{ ami.image_id }}"
    region: "{{ region }}"
    security_groups: "{{ security_group_web.group_name }}"
    instance_type: "{{ ec2.instance_type }}"
    assign_public_ip: "{{ ec2.public_ip }}"
    user_data: "{{ lookup('file', tmp_file_create_cloud_init_web_task.
path) }}"
```

As you can see, this uses the `community.aws.autoscaling_launch_config` module as there is currently no official support in the `amazon.aws` collection for the creation of launch configurations.

The final task in the role, and also the final one where we will target AWS directly, is as follows:

```
- name: Create the Auto Scaling Group
  amazon.aws.autoscaling_group:
    name: "{{ asg_name }}"
    region: "{{ region }}"
    target_group_arns:
      - "{{ alb_target_group.target_group_arn }}"
    availability_zones:
      - "{{ zones.availability_zones[0].zone_name }}"
      - "{{ zones.availability_zones[1].zone_name }}"
    launch_config_name: "{{ lauch_configuration_name }}"
    min_size: "{{ ec2.asg.min_size }}"
    max_size: "{{ ec2.asg.max_size }}"
    desired_capacity: "{{ ec2.asg.desired_capacity }}"
    vpc_zone_identifier:
      - "{{ subnet_web01.subnet.id }}"
      - "{{ subnet_web02.subnet.id }}"
    tags:
      - Name: "{{ asg_name }}"
      - Description: "{{ dict.ansible_warning }}"
      - Project: "{{ app.name }}"
      - Environment: "{{ app.env }}"
      - Deployed_by: "Ansible"
```

This creates the ASG, which will immediately start to launch however many instances we have defined in the `{{ ec2.asg.desired_capacity }}` variable.

All values are again filled using hardcoded variables like the one we just mentioned or output registers.

The Output role

All that is left now is to print some information to the terminal containing the URL we need to open to visit our WordPress installation and the credentials needed to log in:

```
- name: Output details on the deployment
  ansible.builtin.debug:
    msg:
      - "Wordpress Admin Username: {{ wordpress.wp_admin_user }}"
      - "Wordpress Admin Password: {{ wp_password }}"
      - "Wordpress URL: http://{{ alb.dns_name }}/"
```

That concludes our Ansible playbook, so let's look at executing it.

Running the Ansible playbook

The command to run the playbook is the same as we used in *Chapter 4, Deploying to Microsoft Azure*:

```
$ ansible-playbook site.yml
```

Once finished, you should see something like the following output:

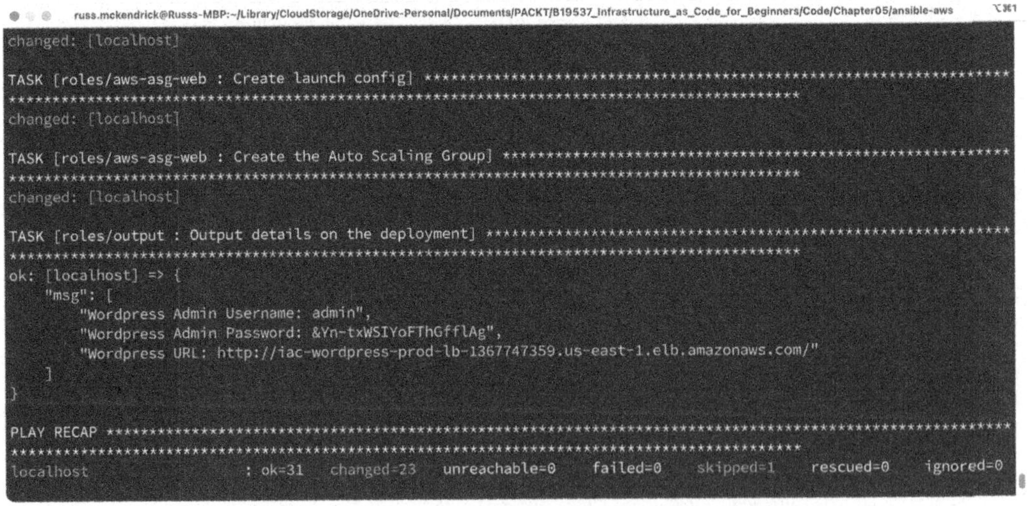

Figure 5.4 – The last few lines of the playbook run output

If you were watching the output, you might have noticed where we put in the logic to wait for the admin EC2 instances to enter the running state. Those lines can be found in the following screenshot:

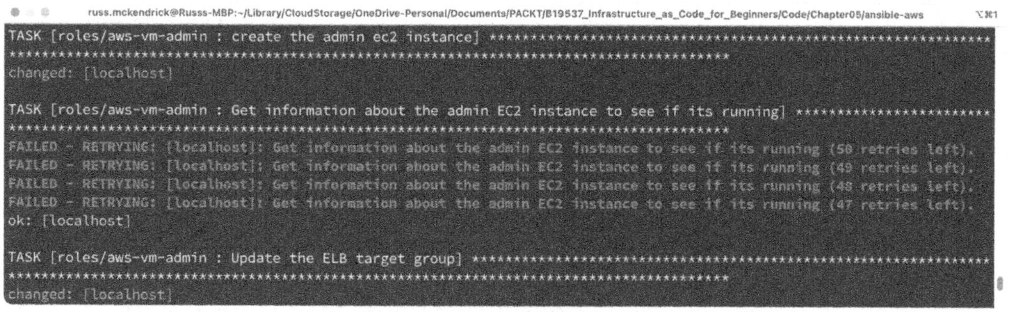

Figure 5.5 – Waiting for the instance to have a state of running

You can now follow the URL in the output and take a look at your WordPress installation. It should look like the Azure installation did in *Chapter 4, Deploying to Microsoft Azure*, and the AWS resources in the AWS Management Console at `http://console.aws.amazon.com`.

Once you are finished looking around, you can terminate all of the resources launched by the playbook by running the following command:

```
$ ansible-playbook destory.yml
```

You may notice that a lot more is happening, as seen in the following output:

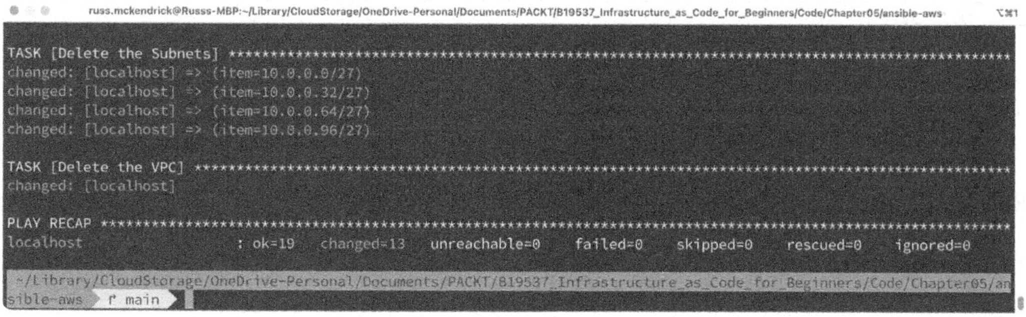

Figure 5.6 – Deleting all of the resources

In fact, there are nearly 20 tasks compared to the small handful when we ran the same playbook in *Chapter 4, Deploying to Microsoft Azure*; why is that?

This is another difference between Microsoft Azure and AWS. When we deployed the resources in Microsoft Azure, we deployed them to a single resource group, which acts as a logical container for your workload, collecting all its resources together.

When we came to terminate our Microsoft Azure deployment, we had to remove the resource group and all the resources contained within it in a single task.

However, AWS is very different, and we need to build a playbook to terminate the resources in the reverse order in which we deployed them.

Some of the tasks used in the destroy.yml file reuse a little of the logic we used in the roles to deploy the resources, so before we look at using Terraform in AWS, let's quickly discuss the destroy.yml playbook, starting with the Auto Scaling group which will remove the instances we have launched.

Auto Scaling group

There are three tasks that deal with removing the ASG; the first task uses the amazon.aws.autoscaling_group_info module to get information on the ASG.

The second task uses the amazon.aws.autoscaling_group module with just enough configuration to allow us to set state to absent – but only when there are more than 0 results returned from the previous task. To do this, we use the following line:

```
when: asgs.results | length > 0
```

This means that the task will be skipped if the ASG has been removed, but we need to rerun the playbook because of another failed task. We will be using this logic throughout this playbook.

The final of the three tasks removes the launch configuration using the community.aws.autoscaling_launch_config module.

EC2 instance

Just the two tasks are required here, one that uses amazon.aws.ec2_instance_info to get information on our EC2 instances, and the second that uses amazon.aws.ec2_instance to set state to absent if the first task returns a result.

RDS instance

There are three tasks here. The first gets information, the second terminates the RDS instance, and the third removes the RDS subnet group.

EFS instance

Just a single task is required here; it uses community.aws.efs to ensure that any resources matching {{ efs_name }} in the region defined by {{ region }} are absent.

Elastic load balancer

Here we have two more simple tasks that use amazon.aws.elb_application_lb and community.aws.elb_target_group to set our state resource to absent.

Security groups

If you remember, when we added the security groups, we used the ID of the Web security group to allow access to the RDS and EFS resources. Also, as we discussed when launching the EC2 instance, the Ansible playbook can sometimes be a little ahead of the AWS API when it comes to completing tasks, meaning that Ansible could be trying to move on to the next task before AWS has completed processing an earlier task.

Because of this, there is the risk that either the RDS or EFS security group may not be fully removed before the playbook attempts to remove the web security group, which would result in a dependency error.

To avoid this, we have a little checking built into the task:

```
- name: Delete the security groups
  amazon.aws.ec2_security_group:
    name: "{{ item }}"
    region: "{{ region }}"
    state: absent
  with_items:
    - "{{ vpc_security_group_name_efs }}"
    - "{{ vpc_security_group_name_rds }}"
    - "{{ vpc_security_group_name_web }}"
  register: delelte_security_groups
  until: "delelte_security_groups is not failed"
  retries: 25
  delay: 10
```

As you can see, we are using `with_items` to loop through our three security groups and set their `state` to `absent`. We also have an `until` set, which will repeat whichever part of the loop fails until it has successfully removed the security group:

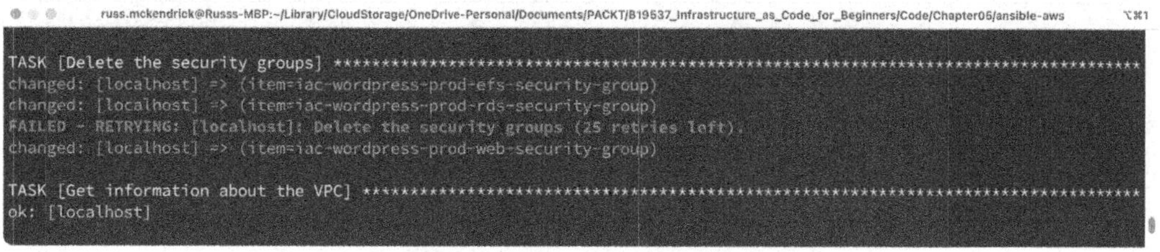

Figure 5.7 – Setting the state of the security groups to absent

It will allow for 25 failures and will try every 10 seconds. As you can see from the preceding screenshot, it should only fail once or twice before moving on.

Virtual Private Cloud

The remaining tasks all work in the same pattern as we defined previously, apart from the route table. Like other resources, we use a module, `amazon.aws.ec2_vpc_route_table_info` in this case, to get information on the route tables. However, the difference here is that it will return the default route table, which was created when we first launched the VPC. This one will error if we try to remove it.

To get around this, we have to extend the when clause in the task:

```
- name: Delete the Route Table
  amazon.aws.ec2_vpc_route_table:
    route_table_id: "{{ item.route_table_id }}"
    vpc_id: "{{ the_vpc.vpcs[0].id }}"
    region: "{{ region }}"
    lookup: id
    state: absent
    when: the_vpc.vpcs | length > 0 and item.associations[0].main !=
true
    with_items: "{{ the_route_tables.route_tables }}"
```

As you can see, this will remove the route table if there are more than zero of them listed and it is not the `main` association. This looks something like the following when run:

Figure 5.8 – Removing our custom route table but skipping the main one

The remaining tasks follow the same patterns we used elsewhere in the chapter when we launched the resources.

> **Information**
>
> Remember to make sure that all the resources have been removed by checking to see whether they are still listed in the AWS Management Console at `http://console.aws.amazon. com`, as you could incur unexpected costs if the preceding playbook failed for any reason.

That is the end of the playbook, which removes the resources and concludes our deep dive into running Ansible on AWS.

Now it is time to move on to Terraform.

Terraform – reviewing the code and deploying our infrastructure

As we did a deep dive into Terraform in *Chapter 4, Deploying to Microsoft Azure*, we aren't going to dig too deep into the code here, and instead will just highlight any considerations we need to make when targeting AWS or if there is a function we didn't use when deploying our workload to Microsoft Azure.

Walk-through of Terraform files

What follows is a walk-through of each of the Terraform files. Just as we did for Microsoft Azure, I have grouped each logical group of resources in its own `.tf` file.

Setup

This is not too dissimilar to the one we defined for Azure. There are a few obvious differences – the biggest of which is that we are using the AWS provider:

```
aws = {
    source  = "hashicorp/aws"
    version = "~> 4.0"
}
```

Also, we are hardcoding the region we want to launch our resources in as a provider configuration option:

```
provider "aws" {
  region = "us-east-1"
}
```

There are a few omissions in that we are not loading any helper providers or modules to assist us with resource naming; that is going to be up to us to define as we launch our resources.

Networking

There are several tasks here:

- `resource "aws_vpc" "vpc"`, which launches the VPC
- `resource "aws_subnet" "web01"`, which adds the web01 subnet
- `resource "aws_subnet" "web02"`, which adds the web02 subnet
- `resource "aws_subnet" "rds01"`, which adds the rds01 subnet
- `resource "aws_subnet" "rds02"`, which adds the rds02 subnet

The four subnet tasks all look similar:

```
resource "aws_subnet" "web01" {
  vpc_id             = aws_vpc.vpc.id
  cidr_block         = cidrsubnet("${aws_vpc.vpc.cidr_block}", 3, 0)
  availability_zone  = var.zones[0]
  tags               = merge(var.default_tags, tomap({ Name = "${var.
name}-${var.environment_type}-web01-subnet" }))
}
```

As you can see, there is a slight difference in the way we are defining the CIDR range for each subnet; rather than hardcode it as we did for Microsoft Azure, we are following a similar pattern as we did when using Ansible and are using a Terraform function called `cidrsubnet` to generate the correct CIDR range for us.

The only other to note is that we are taking the list of `default_tags` we are defining in our `tfvars` file and merging it with a map we are dynamically creating using the `tomap` function. This map name contains the Name tag. We will be reusing this approach throughout the remainder of the deployment.

The remaining tasks are very similar to the ones we performed when deploying using Ansible:

- `resource "aws_internet_gateway" "vpc_igw"`, which deploys an internet gateway.
- `resource "aws_route_table" "vpc_igw_route"`, which adds a route table to route all outgoing traffic to the internet gateway.
- `resource "aws_route_table_association" "rta_subnet_public01"`, which associates the route table we just created with the web01 subnet.
- `resource "aws_route_table_association" "rta_subnet_public02"`, which associates the route table we just created with the web02 subnet.
- `resource "aws_security_group" "sg_vms"`, which creates a security group opening ports 80 and 22 to everyone, that is, 0.0.0.0/0.

- resource "aws_security_group" "sg_efs", which adds the EFS security group opening port 2049 to any resource with the web security group attached.

- resource "aws_security_group" "sg_rds", which creates the RDS security group opening port 3306 to any resource with the web security group attached.

- resource "aws_lb" "lb", which creates an elastic load balancer with a type of application.

- resource "aws_lb_target_group" "front_end", which creates the target group we will be registering our EC2 instances with.

- resource "aws_lb_listener" "front_end", which configures the frontend listener for port 80 on the elastic load balancer. When we launched our workload using Ansible, this was defined in line when creating the ELB resource.

That is all the network resources we need to launch and configure to support the remaining services for our workload. Now we can start defining the resources themselves.

Storage

There are three tasks in this file, which are as follows:

- resource "aws_efs_file_system" "efs", which creates the Amazon EFS volume

- resource "aws_efs_mount_target" "efs_mount_targets01", which creates the mount target in the web01 subnet

- resource "aws_efs_mount_target" "efs_mount_targets02", which creates the mount target in the web02 subnet

Now that our storage is in place, we can move into the Amazon RDS instance.

Database

Again, we have just three tasks defined here to configure and launch our Amazon RDS instance. They are as follows:

- resource "aws_db_subnet_group" "database", which creates the subnet group so that our Amazon RDS instance is accessible from within our VPC

- resource "random_password" "database_password", which randomly generates the password we will be using when launching the Amazon RDS service

- resource "aws_db_instance" "database", which deploys the Amazon RDS instance into the subnet group we defined and configures it per the variables we have defined in the variables file

As you already know, as we are following the steps we took when launching the workload with Ansible, it is now time to launch the Admin EC2 instance and bootstrap WordPress.

Virtual machine (admin)

First of all, we need to find the right AMI to use. This slightly differs from how we achieved this using Ansible, as Terraform can pick the latest one for us as part of the task execution:

```
data "aws_ami" "ubuntu_admin" {
  most_recent = var.ami_most_recent
  owners      = [var.ami_owners]
  filter {
    name   = "name"
    values = [var.ami_filter_name]
  }
  filter {
    name   = "virtualization-type"
    values = [var.ami_filter_virtualization_type]
  }
}
```

As you can see, we are setting the `most_recent` key to be the value of the `var.ami_most_recent` variable, which by default is set to `true`.

Before we launch the EC2 instance, we have one last bit of housekeeping to do, and that is to create the WordPress admin password:

```
resource "random_password" "wordpress_admin_password" {
  length           = 16
  special          = true
  override_special = "_%@"
}
```

Now we have everything we need to launch our EC2 instance. To start, we define the basics needed to launch the instance:

```
resource "aws_instance" "admin" {
  ami = data.aws_ami.ubuntu_admin.id
  instance_type = var.instance_type
  subnet_id = aws_subnet.web01.id
  associate_public_ip_address = true
  availability_zone = var.zones[0]
  vpc_security_group_ids = [aws_security_group.sg_vms.id]
```

The next part of the task is where the user data is defined. More on that in a second:

```
user_data = templatefile("vm-cloud-init-admin.yml.tftpl", {
    tmpl_database_username = "${var.database_username}"
    tmpl_database_password = "${random_password.database_password.
result}"
    tmpl_database_hostname = "${aws_db_instance.database.address}"
    tmpl_database_name     = "${var.database_name}"
    tmpl_file_share        = "${aws_efs_file_system.efs.dns_name}"
    tmpl_wordpress_url     = "http://${aws_lb.lb.dns_name}/"
    tmpl_wp_title          = "${var.wp_title}"
    tmpl_wp_admin_user     = "${var.wp_admin_user}"
    tmpl_wp_admin_password = "${random_password.wordpress_admin_
password.result}"
    tmpl_wp_admin_email    = "${var.wp_admin_email}"
})
```

Finally, we define the tags, which include the resource name:

```
tags = merge(var.default_tags, tomap({ Name = "${var.name}-${var.
environment_type}-ec2-admin" }))
}
```

As you can see, we are using a similar logic to when we launched the workload in Microsoft Azure when injecting `user_data` by using the `templatefile` function. However, we are not having to Base64 encode it this time around.

The template for the `cloud-init` file contains the same changes we made when launching the workload using Ansible, again using `nc` to check that the DNS endpoint for the NFS share is responding on port `2048` before mounting the volume. The only other differences are related to the templating functions between the two tools.

The final task, like Ansible, is to register the newly launched EC2 instance with our ELB target group:

```
resource "aws_lb_target_group_attachment" "admin" {
  target_group_arn = aws_lb_target_group.front_end.arn
  target_id        = aws_instance.admin.id
  port             = 80
}
```

The final difference between using Terraform and Ansible is that we do not have to build the logic to wait for the EC2 instance to have a state of `running` in our code, as Terraform continues to poll the state of the EC2 instance until it is as desired – which by default is `running`.

Any dependencies on the EC2 task, like our `"aws_lb_target_group_attachment"` `"admin"` task, will not error as the Ansible deployment did because the deployment won't progress until the state is met.

Auto Scaling group (web)

As with Ansible, the final set of AWS resources we will launch is the ASG for the web servers.

Again, we start with a launch configuration:

```
resource "aws_launch_configuration" "web_launch_configuration" {
  name_prefix                 = "${var.name}-${var.environment_type}-
alc-web-"
  image_id                    = data.aws_ami.ubuntu_admin.id
  instance_type               = var.instance_type
  associate_public_ip_address = true
  security_groups             = [aws_security_group.sg_vms.id]
  user_data = templatefile("vm-cloud-init-web.yml.tftpl", {
    tmpl_file_share = "${aws_efs_file_system.efs.dns_name}"
  })
}
```

Like when we deployed in Microsoft Azure, the only variable we needed to pass into the template file was the DNS endpoint for the Amazon EFS endpoint.

Now that the launch configuration is in place, we can create the ASG, which will immediately launch the number of EC2 instances we have defined in the var.min_number_of_web_servers variable:

```
resource "aws_autoscaling_group" "web_autoscaling_group" {
  name                 = "${var.name}-${var.environment_type}-asg-web"
  min_size             = var.min_number_of_web_servers
  max_size             = var.max_number_of_web_servers
  launch_configuration = aws_launch_configuration.web_launch_
configuration.name
  target_group_arns    = [aws_lb_target_group.front_end.arn]
  vpc_zone_identifier  = [aws_subnet.web01.id, aws_subnet.web02.id]

  lifecycle {
    create_before_destroy = true
  }
}
```

With that task in place, we have everything needed to launch the workload, apart from the output, which tells us how to access WordPress.

Output

There are three outputs defined here – one of which is being marked as sensitive:

```
output "wp_user" {
  value    = "Wordpress Admin Username: ${var.wp_admin_user}"
```

```
    sensitive = false
}
output "wp_password" {
    value     = "Wordpress Admin Password: ${random_password.wordpress_
admin_password.result}"
    sensitive = true
}
output "wp_url" {
    value     = "Wordpress URL: http://${aws_lb.lb.dns_name}/"
    sensitive = false
}
```

This will give you the URL you can use to access your WordPress site, along with the username and password. Now we can run our Terraform script.

Deploying the environment

To deploy the environment, we simply need to run the following commands:

```
$ terraform init
$ terraform apply
```

Answering yes when prompted after running terraform apply will proceed with the deployment, and once complete, you should see something like the following screen:

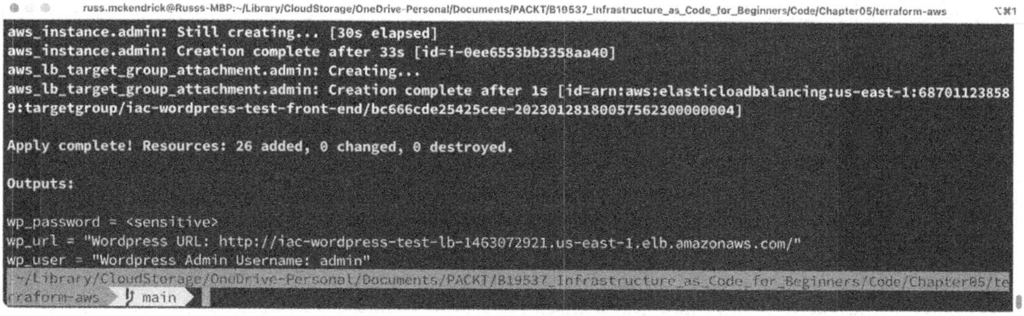

```
 ● ● ●      russ.mckendrick@Russ-MBP:~/Library/CloudStorage/OneDrive-Personal/Documents/PACKT/B19537_Infrastructure_as_Code_for_Beginners/Code/Chapter05/terraform-aws     ⌥⌘1
aws_instance.admin: Still creating... [30s elapsed]
aws_instance.admin: Creation complete after 33s [id=i-0ee6553bb3358aa40]
aws_lb_target_group_attachment.admin: Creating...
aws_lb_target_group_attachment.admin: Creation complete after 1s [id=arn:aws:elasticloadbalancing:us-east-1:68701123858
9:targetgroup/iac-wordpress-test-front-end/bc666cde25425cee-20230128180057562300000004]

Apply complete! Resources: 26 added, 0 changed, 0 destroyed.

Outputs:

wp_password = <sensitive>
wp_url = "Wordpress URL: http://iac-wordpress-test-lb-1463072921.us-east-1.elb.amazonaws.com/"
wp_user = "Wordpress Admin Username: admin"
 ~/Library/CloudStorage/OneDrive-Personal/Documents/PACKT/B19537_Infrastructure_as_Code_for_Beginners/Code/Chapter05/te
rraform-aws   ⎇ main
```

Figure 5.9 – Deploying the environment using Terraform

Again, like we did when deploying to Microsoft Azure, running terraform output -json will show the content of the sensitive value, meaning you can browse and log in to WordPress and review the resources in the AWS Management Console.

When you have finished, you just need to run the following command:

```
$ terraform destroy
```

This will remove all the resources that we created using the `terraform apply` command. As always, double-check in the AWS Management Console that all of the resources have been removed correctly, as you do not want to incur any unexpected costs.

Summary

In this chapter, we have done a deep dive into using Ansible to deploy our WordPress environment in AWS.

After discussing what our deployment looks like, we walked through the Ansible playbook and expanded on the quick overview that we had in *Chapter 4, Deploying to Microsoft Azure*. We discussed Ansible roles and how to bootstrap one using the `ansible-galaxy init` command.

We discussed some of the built-in functions and utilities, such as `ipsubnet`, `sort`, and `regex_replace`, which we used to manipulate hardcoded and output variables. We also covered a few different approaches for building logic into our playbook tasks using functions such as `until` to make sure that our playbook does not error both when launching the resources and, just as importantly, when terminating resources. After all, we don't want stray resources hanging around and costing money.

We then took a quick look at how we would deploy the same resources using Terraform, as we had already done a deeper dive into Terraform, highlighting some additional approaches we can take when deploying resources.

Throughout both walk-throughs, we also discussed the differences in the approach we needed to take to deploy our workload in AWS as opposed to Microsoft Azure.

Feel free to play around with both the Ansible and Terraform code; for example, try and update the number of servers being launched, update the various SKUs, change the network addressing, and so on, and see what effects your changes have on the deployment.

In the next chapter, we will expand on what we have covered in this chapter and *Chapter 4, Deploying to Microsoft Azure*, by looking further at how the two cloud-agnostic tools we have been looking at work and what considerations we need to make when approaching the cloud providers.

We will also examine how we can make our Ansible and Terraform code more reusable.

Further reading

You can find more details on the services and documentation we have mentioned in this chapter at the following URLs:

- Amazon services:

  - Amazon ELB: `https://aws.amazon.com/elasticloadbalancing/`

  - Amazon EC2: `https://aws.amazon.com/ec2/`

  - Amazon EFS: `https://aws.amazon.com/efs/`

  - Amazon RDS: `https://aws.amazon.com/rds/`

  - Amazon VPC: `https://aws.amazon.com/vpc/`

  - Amazon EC2 ASGs: `https://docs.aws.amazon.com/autoscaling/ec2/userguide/auto-scaling-groups.html`

- Ansible collections:

  - Amazon AWS collection: `https://galaxy.ansible.com/amazon/aws`

  - Community AWS collection: `https://galaxy.ansible.com/community/aws`

- Terraform provider:

  - HashiCorp AWS: `https://registry.terraform.io/providers/hashicorp/aws/`

6

Building upon the Foundations

As we continue to navigate the ever-evolving landscape of cloud computing, it is essential to understand the nuances of deploying high-level designs across different public cloud providers. In this chapter, we will investigate the differences that arise when using cloud-agnostic tools such as Terraform and Ansible.

I have found that despite our best efforts to maintain consistency, variations will always crop up when deploying designs across different providers. In this chapter, I will share some of my own experiences addressing these variations and provide some practical approaches for building repeatable deployment processes for various applications and environments.

We will also discuss the importance of creating modular code, a crucial aspect of streamlining deployment efforts and avoiding duplicating code. By implementing these techniques, we can efficiently and effectively deploy our designs across different public cloud providers.

We will cover the following topics in the chapter:

- Understanding cloud-agnostic tools
- Understanding the differences between our two cloud deployments
- Understanding the differences between our Terraform and Ansible deployments
- Introducing more variables
- Making the code more reusable

Let's start by discussing how cloud-agnostic the tools we have been using are.

Understanding cloud-agnostic tools

In *Chapter 4, Deploying to Microsoft Azure,* and *Chapter 5, Deploying to Amazon Web Services,* we use both Terraform and Ansible to target these clouds – so we know they work with both cloud providers, but how much of the code did we reuse?

The honest answer is very little.

We used different providers/collections for each of the cloud providers. As a result, there were many allowances we needed to make. While conceptually, the cloud providers offer like-for-like services at a high level; they have evolved in very different ways to achieve the same task.

For example, launching something as simple as a virtual machine requires two approaches: deploying services such as networking work requires different considerations and configurations as they simply just work differently.

So why do we call the two tools we have been looking at cloud agnostic? Surely that should mean *they just work*.

In an ideal world, yes, that should be the case. With the trends in AI tools at the time of writing this book in early 2023, we might be close enough to define our **Infrastructure-as-Code** (**IaC**) deployments in natural language, with some constraints and rules, and have it target our cloud of choice.

While that may be close, it doesn't exist now.

So back to the here and now with the two tools we have been working with, what changes could we make to how we work to make them as cloud agnostic as possible?

As we discovered in *Chapter 4, Deploying to Microsoft Azure*, and *Chapter 5, Deploying to Amazon Web Services*, both Terraform and Ansible have some useful helper functions, tools, and utilities, so the more we can take advantage of these across our deployments, the better.

Throughout the rest of this chapter, we will look at what we can consistently use across our deployments, no matter which cloud we are targeting.

To do this, we need to look at the consistencies across our deployment and then figure out how we can best take them into account in our deployments by developing a more standard approach to writing, managing, and executing our code.

Understand the differences between our Microsoft Azure and Amazon Web Services deployments

Let's summarize our deployment as we have covered the deployment in four separate sets of code across the previous two chapters.

General

There is just a single service here, and as you can see – it is only available in one of our target cloud providers:

Service/Function	Microsoft Azure	Amazon Web Services
Resource Container	Resource group	Not available

There isn't an equivalent of resource groups within Amazon Web Services, though some could argue that tagging does the same job. However, tags act more as a way of searching for and reporting against your resources rather than collecting them all together in a container, which, as we have seen, can be removed or have permissions applied to them.

Network

Next up, we have the network resources; any resources marked with a *, while available, are not used in our WordPress deployment:

Service / Function	Microsoft Azure	Amazon Web Services
Network	Virtual network	**Virtual Private Cloud (VPC)**
Subnet	Subnet	Subnet
Gateway	NAT gateway *	Internet gateway
Route table	Route table *	Route table
Security	Network security groups	Security group
Load balancer	Load Balancer/Application Gateway *	Elastic Load Balancer */Application Load Balancer

From a service point of view, we have an even coverage of services. At the same time, they are configured slightly differently between the two cloud services:

- **Resource name**: All Azure resources require a name.
- **Resource regions and availability zones**: Both clouds have a concept of regions – and in most of those regions, there are multiple availability zones though it is worth pointing out that some secondary regions in Microsoft Azure – for example, UK West, do not have availability zones.
- **Classless Inter-Domain Routing (CIDR) range**: The networks need a CIDR range; in our example, this was 10.0.0.0/24.

- **Subnet addresses**: There were some critical differences between the subnets deployed in our two clouds; for example, in Microsoft Azure, we needed to delegate a particular service to them, whereas, in AWS, we didn't need to delegate a service. Still, we did pin our subnets to an availability zone within our target region. However, outside of this, the information needed for each cloud is roughly the same.

Of the network services we are deploying, a few of them would benefit from being configured using loops and passing in variables – though this could get a little complicated as we will need a little logic for both Terraform and Ansible for Azure Services, which may require a service delegated to the subnet.

Storage

On the face of it, this should be simple as we *just* need to launch and configure some storage; however, as you may remember from our scripts, there are pretty big differences between the two cloud providers in terms of storage:

Service/Function	Microsoft Azure	Amazon Web Services
Storage (**Network File System (NFS)**)	Storage account with Azure Files enabled	Amazon Elastic File Service
Private **Domain Name System (DNS)**	Private DNS zone	Mount targets
Network integration	Private endpoint	Not required

As you can see, there are some differences in the way that Microsoft handles network integration of its services in Azure – with the key word there being **integration**.

The most significant and consistent difference between the two cloud providers is how networking works on their **Platform-as-a-Service (PaaS)** services.

I typically explain that Amazon has built its PaaS services from the ground up to be deployed within an Amazon VPC network.

By contrast, Microsoft has built its PaaS services to allow you to link them to your virtual network. In some cases, that link is not always bi-directional, so certain PaaS services can only have access to resources within a virtual network rather than being able to be consumed within the virtual network – while this is not the case for any of the services in our example WordPress deployment, it is a consideration you need to make when planning your deployments.

The information required to launch and configure the services is similar, even with the differences previously described.

Database

In typical fashion, after explaining that in Microsoft Azure, most PaaS services have a level of virtual network integration rather than being launched directly into the network, we launch one of the network Azure services that is hosted within the virtual network:

Service/Function	Microsoft Azure	Amazon Web Services
Database	Azure Database for MySQL – Flexible Server	Amazon Relational Database Service
Private DNS	Private DNS zone	Subnet group

While we don't need to add a private endpoint when deploying Azure Database for MySQL – Flexible Server, we do need to delegate an entire subnet to the service, so there are a few considerations still to make when planning the deployment.

Again, the bulk of the information required to launch the services is similar between the two cloud providers.

Virtual machine (admin)

When deploying the admin virtual machine instance, we needed to make a few considerations; however, the information required is similar for each of our two cloud providers:

Service/Function	Microsoft Azure	Amazon Web Services
Image	Azure Image from the Azure Marketplace	Amazon Machine Image (AMI) from the AWS Marketplace
Compute	Azure Virtual Machine	Amazon Elastic Compute Cloud (EC2)
Load Balancer Attachment	Required	Required

As you may recall, when we launched our WordPress workload in Amazon Web Services, we needed to adjust our cloud-init script slightly to consider the differences in how some of the services are consumed. All we needed was to build in a bit of logic to check and, if required, wait for our resources to be available.

Virtual machines with scaling (web)

Everything we mentioned for deploying the admin virtual machine instance also applies here; there is only really one main difference between the two providers:

Service/Function	Microsoft Azure	Amazon Web Services
Image	Azure Image from the Azure Marketplace	AMI from the AWS Marketplace
Configuration	Not required	Launch configuration
Compute	Azure Virtual Machine Scale Sets	Amazon EC2 Auto Scaling Group
Load Balancer Attachment	Inline	Inline

As you can see, all Azure configurations are inline; however, by contrast, Auto Scaling groups in Amazon Web Services require a launch configuration to use as the base for our deployment.

Seeing it in action

As you can see, while the two cloud providers work slightly differently, there is close-enough feature parity for your deployment to take a similar approach, at least at a high level.

So, what does all this mean when it comes to being cloud agnostic with a single tool?

Well, as we have already discussed at a high level, the approach is similar, and while the modules/tasks may differ, you can use some of the same logic when it comes to your deployments.

Let's look at doing this with Terraform code; the code will create a primary network and then use a loop to create four subnets in both Microsoft Azure and Amazon Web Services:

1. First of all, let's look at the variables we are going to be using to achieve this – to start with, we have the `name`, `region`, and `default` tags:

```
variable "name" {
  description = "Base name for resources"
  type        = string
  default     = "iac-wordpress"
}
variable "region" {
  description = "The region to deploy to"
  type        = string
  default     = "uksouth"
```

```
}
variable "tags" {
  description = "The default tags to use across all of our
resources"
  type        = map(any)
  default = {
    project     = "iac-wordpress"
    environment = "example"
    deployed_by = "terraform"
  }
}
```

The only variable that will change between our two cloud providers is `region`, as each provider has different region names.

2. Next up, we define the address space:

```
variable "address_space" {
  description = "The address space of the network"
  type        = string
  default     = "10.0.0.0/24"
}
```

3. Nothing too special here still; however, for the subnets, we define the following, from which, while quite lengthy, you should be able to quickly get an idea of what is happening:

```
variable "subnets" {
  description = "The subnets to deploy the network"
  type = map(object({
    name                  = string
    address_prefix_size   = number
    address_prefix_number = number
  }))
  default = {
    subnet_001 = {
      name                  = "subnet001"
      address_prefix_size   = "3"
      address_prefix_number = "0"
    },
    subnet_002 = {
      name                  = "subnet002"
      address_prefix_size   = "3"
      address_prefix_number = "1"
    },
    subnet_003 = {
      name                  = "subnet003"
      address_prefix_size   = "3"
```

```
            address_prefix_number = "2"
        },
        subnet_004 = {
          name                   = "subnet004"
          address_prefix_size    = "3"
          address_prefix_number  = "3"
        },
      }
    }
```

As you can see, we are defining a map here as that will give us something we can loop through. Now let us move on to the main.tf files.

4. We first create the network itself; the following is for AWS, where we are creating a VPC:

```
resource "aws_vpc" "network" {
  cidr_block            = var.address_space
  tags                  = merge(var.tags, tomap({ Name = "${var.
name}-vpc" }))
}
```

5. Now we have the same task, but this time for Azure, which creates a virtual network:

```
resource "azurerm_virtual_network" "network" {
  resource_group_name = azurerm_resource_group.resource_group.
name
  location            = azurerm_resource_group.resource_group.
location
  name                = "vnet-${var.name}-${var.region}"
  address_space       = [var.address_space]
  tags                = merge(var.tags, tomap({ Name = "vnet-
${var.name}-${var.region}" }))
}
```

As you can see, they are not too dissimilar, and we are applying the same logic of taking the list of tags and adding one using the merge function to add the resource name.

6. Now that we have our networks, it's time to loop over the subnets variable and add those, starting with AWS again:

```
resource "aws_subnet" "subnets" {
  for_each              = var.subnets
  vpc_id                = aws_vpc.network.id
```

```
    cidr_block            = cidrsubnet("${aws_vpc.network.cidr_
block}", each.value.address_prefix_size, each.value.address_
prefix_number)
    tags                  = merge(var.tags, tomap({ Name = "${var.
name}-${each.value.name}" }))
}
```

7. Then the same again, this time for Azure:

```
resource "azurerm_subnet" "subnets" {
  for_each              = var.subnets
  name                  = each.value.name
  resource_group_name   = azurerm_resource_group.resource_group.
name
  virtual_network_name = azurerm_virtual_network.network.name
  address_prefixes      = [cidrsubnet("${azurerm_virtual_network.
network.address_space[0]}", each.value.address_prefix_size,
each.value.address_prefix_number)]
}
```

As you can see, we are using the same approach in both in that we are looping through the var.subnets variable using a for_each loop.

We then use each.value.name to name the resource, in Azure's case, using the name key, and for AWS, by creating a Name tag.

For both, we use the output of creating the network to reference it; for AWS, we use aws_vpc.network.id; in Azure, we use azurerm_virtual_network.network.name.

This will ensure that Terraform will only attempt to create the subnets once the network they are going to live in has been created.

We can then use the cidrsubnet function to take our address space, which again is being referenced from the network resource we created using "${aws_vpc.network.cidr_block}" for AWS and "${azurerm_virtual_network.network.address_space[0]}" for Azure.

We then use each.value.address_prefix_size to define the CIDR size of each subnet, which in our case is /27, and each.value.address_prefix_number to define where within the address space /27 is placed.

As you can see, while the application of variables and functions is slightly different between Amazon Web Services and Microsoft Azure, we can use the same logic to generate and loop through the subnets.

We can also apply the same logic using Ansible – as we will discuss more in the next section.

Understanding the differences between our Terraform and Ansible deployments

We have discussed how we can take a cloud-agnostic approach to our deployments when using either Terraform or Ansible, as each tool has built-in functions and logic for manipulating our variables and the output of running tasks.

Some big differences should have become apparent during the code walk-throughs in *Chapter 4, Deploying to Microsoft Azure,* and *Chapter 5, Deploying to Amazon Web Services.* I am also sure you are forming an opinion on which of the two tools you prefer.

The two tools are very different in their approach, which is to be expected as they were designed to do two different tasks.

Terraform is primarily designed to manage infrastructure, whereas Ansible manages server and state configuration, which also includes a level of infrastructure management.

During my day job, I have, and continue to use, both tools – so where does the decision to use one or the other come in?

If a project requires the repeatable deployment and configuration of several PaaS services in either cloud – especially if the resources need to be launched, consumed, and then terminated, then I recommend using Terraform; this is for a few reasons:

- First, it stores everything in its state file, making terminating any workload a lot more straightforward as we discovered when terminating our AWS deployment using Ansible, we needed to build in quite a lot of logic to make sure that the workload was correctly terminated and removed.

- Secondly, it plays well into **continuous integration/continuous delivery (CI/CD)** services such as GitHub Actions, which we will discuss in more detail in *Chapter 7, Leveraging CI/CD in the Cloud.*

- Finally, I find it has much more coverage and support for some of the newer services and features the cloud providers are introducing. That's not to knock the Ansible development team; it is just that Ansible, in most cases, appears to lag Terraform with new features depending on which of the clouds you are targeting.

Some of the reasons to use Terraform are also some of the contributing factors when it comes to choosing to use Ansible – for example, as Ansible does not use state files and dynamically discovers resources, it is a lot more straightforward to manage in-life changes, for instance, ones that are made once the resources have been deployed and the service is in production, as you don't run the risk of the tool trying to enforce a state it knows about strictly.

Also, Ansible can be used if I need to interact with the host at the resource level itself, for example, I need to, **Secure Shell (SSH)** into a server that has just been launched or target a Windows Server using WinRM to configure the host to set Apache or **Internet Information Services (IIS)**.

It is great for working with fixed points, meaning that, let's say, you have been using Ansible to manage the state of the workload, which has been running a virtual machine on-premise when you can likely reuse a lot of that code to target a cloud environment.

In these cases, Ansible will be the tool of choice.

There is also another option – *use both*! That's right; you can use Ansible to run your Terraform code using the `community.general.terraform` task.

In the code repository accompanying this title, you will find a folder called `ansible-terraform-azure`. This contains an Ansible playbook, which will use Terraform to launch an Azure-hosted virtual machine and then, using Ansible, connect to it, install **NGINX**, and upload a custom `index.html` file.

The task that performs this is as follows:

```
- name: Launch an Azure Virtal Machine instance and supporting
  resources using Terraform
  community.general.terraform:
    project_path: "./terraform"
    state: "present"
    complex_vars: true
    variables:
      name: "{{ app.name }}"
      region: "{{ azure.region }}"
      address_space: "{{ azure.vnet_address_space }}"
      vm_admin_username: "{{ azure.vm_admin_username }}"
      vm_ssh_public_key: "{{ lookup('file', '{{ ssh.public_key_path
}}') }}"
      tags:
        app: "{{ app.name }}"
        env: "{{ app.env }}"
        deployed_by: "{{ app.deployed_by }}"
    force_init: true
  register: terraform_output
```

As you can see, we are telling the task where our Terraform code is; in this case, it is in the `terrform` folder. We are then passing in several variables, which overwrite the defaults defined in the `variables.tf` file in the `terraform` folder.

As part of the Terraform execution, we are outputting the public IP address and the name of the virtual machine, which we then add to a host group using the following task:

```
- name: Add the Virtual Machine to the vmgroup group
  ansible.builtin.add_host:
    groups: "{{ host_group_name }}"
    hostname: "{{ terraform_output.outputs.vm_name.value }}"
    ansible_host: "{{ terraform_output.outputs.public_ip.value }}"
    ansible_port: "{{ ssh.port_number }}"
```

Before finally setting some facts using the `ansible.builtin.set_fact` module:

```
- name: set some facts based on the virtual machine we just launched
  using Terraform
  ansible.builtin.set_fact:
    ansible_ssh_private_key_file: "{{ ssh.private_key_path }}"
    ansible_ssh_user: "{{ azure.vm_admin_username }}"
    the_public_ip: "{{ terraform_output.outputs.public_ip.value }}"
    the_vm_name: "{{ terraform_output.outputs.vm_name.value }}"
```

If you run the playbook, which you can do by running the following command:

```
$ ansible-playbook -i inv site.yml
```

You should see something like the following output:

Figure 6.1 – Having Ansible run Terraform

If you follow the link given in your output (the one in the previous screenshot is no longer active), you should be presented with a web page that looks like the following screen:

Figure 6.2 – Having Ansible run Terraform

You can remove everything using the following playbook:

```
$ ansible-playbook -i inv destroy.yml
```

As you may have imaged, as we are using Terraform to manage the Azure resources, the preceding playbook uses Ansible to run `terraform destroy` rather than us having to set each resource to `absent` as we have had to do in previous Ansible playbooks.

Now that we have discussed how to use both Ansible and Terraform together to get the best out of both tools, we need to discuss variables next. As you will have noticed, we have used many variables in all our Ansible and Terraform code, so let's now discuss how we can best use them.

Introducing more variables

Personally, I try to do everything I can use variables rather than hardcoding values into the code itself – while this can take more time when it comes to writing your code, I highly recommend it as both tools we have looked at allow you to override variables at runtime via the command line.

To do this in Terraform, you can use the following flag when running the `apply` command:

```
$ terraform apply -var region="eu-west-1"
```

When running the Terraform code, which we discussed earlier in the chapter, we launched a network in Amazon Web Services to change the region from `us-east-1`, which is the default set within the code, to `eu-west-1` instead.

You can add multiple variables; the following example expands on the previous one by adding a new address space to use:

```
$ terraform apply -var region="eu-west-1" -var address_
space="172.16.0.0/24"
```

When running Ansible code, for example, in the playbook we executed in the last section, you can run the following command to override a top-level variable:

```
$ ansible-playbook -i inv site.yml --extra-vars 'host_group_name:
"azure_vms "'
```

Here we are changing the host group name to `azure_vms`; you may have noticed that I said top-level variable because many of the variables we are using in the Ansible playbook are multi-level sub-variables – these look like the following:

```
page:
  title: "Infrastructure as Code for Beginners"
  body: "This Virtual Machine was launched using Terraform and
Ansible"
```

To override these, we need to pass in multiple variables as JSON; this turns our command into the following:

```
$ ansible-playbook -i inv site.yml --extra-vars '{"page": { "title":
"Changing a variable at runtime", "body": "Redeploying using a
different variable" }}'
```

It is important to pass in everything if I try and override a single variable, for example, `title`, as it replaces the whole block, which means that there is no longer a `body` variable, and our playbook run will error.

We will use some of these techniques and more in our next chapter, *Chapter 7, Leveraging CI/CD in the Cloud.*

As mentioned at the start of the section, it can sometimes add more time to creating your code when trying to use as many variables as possible – especially if you have quickly prototyped your code to deploy something and have hardcoded some of the values in for ease of use.

It is possible to use tools such as **ChatGPT** and other AI services to help add variables to take your code and replace hardcoded values quite quickly.

Use the following prompt to ask ChatGPT to turn some basic Terraform code into one with variables:

```
Add variables to this Terraform code using the current values as the
defaults, making sure to add a description and the type of variable.

```

resource "azurerm_resource_group" "example" {
  name   = "example"
  location = "West Europe"
}
```
```

It gave the following response:

```
variable "resource_group_name" {
  description = "The name of the Azure resource group."
  type        = string
  default     = "example"
}

variable "location" {
  description = "The location of the Azure resource group."
  type        = string
  default     = "West Europe"
}

resource "azurerm_resource_group" "example" {
  name     = var.resource_group_name
  location = var.location
}
```

As you can see, we have everything we need to copy and paste into our code – at the time of writing, responses from tools such as ChatGPT need to be double-checked, but as mentioned earlier in this chapter – as tools like this get more powerful, it is fully expected that you will be using them a lot more in your day-to-day workflows.

One of the advantages of using variables is that it makes our code more reusable, let's discuss this in a little more detail now.

Making the code more reusable

As well as using variables, we are also able to reuse *chunks* of code – when we discussed Ansible in *Chapter 5, Deploying to Amazon Web Services*, we discussed *roles*. In Ansible, roles are designed to be called repeatedly, so while we used them to logically split our project into more manageable sections, we can go one step further and have them only perform a single function.

We can also do the same thing in Terraform. For most of our Azure deployments so far, we have been using a module downloaded from the Terraform registry to manage the region settings.

Claranet, the publisher of that module, also has others – let us look at how we can create a virtual network in Azure using only modules (the complete executable code can be found in this book's GitHub repository):

1.   To start, we need to initialize the region module as we have been doing in our other Terraform code:

```
module "azure_region" {
    source        = "claranet/regions/azurerm"
    azure_region = var.region
}
```

2.   Once we have the region locked in, we can then use the output of that module to create a resource group:

```
module "rg" {
    source        = "claranet/rg/azurerm"
    location      = module.azure_region.location
    client_name = var.name
    environment = var.environment
    stack         = var.project_name
}
```

As you can see, we are using `module.azure_region.location` to define the location. Then we are passing in some details about our project – as Claranet is a managed service provider, it uses `client_name` and `stack` throughout its modules.

3.   Next up, we need to create a virtual network:

```
module "azure_virtual_network" {
    source               = "claranet/vnet/azurerm"
    environment          = var.environment
    location             = module.azure_region.location
    location_short       = module.azure_region.location_short
    client_name          = var.name
    stack                = var.project_name
    resource_group_name = module.rg.resource_group_name
    vnet_cidr            = var.address_space
}
```

Again, we can see more of the same information and the CIDR space we want to use.

4.   The final part is to create the subnet(s):

```
module "azure_network_subnet" {
    for_each            = var.subnets
    source              = "claranet/subnet/azurerm"
    environment         = var.environment
    location_short      = module.azure_region.location_short
```

```
        custom_subnet_name    = each.value.name
        client_name           = var.name
        stack                 = var.project_name
        resource_group_name   = module.rg.resource_group_name
        virtual_network_name = module.azure_virtual_network.virtual_
    network_name
        subnet_cidr_list      = [cidrsubnet("${module.azure_virtual_
    network.virtual_network_space[0]}", each.value.address_prefix_
    size, each.value.address_prefix_number)]
        }
```

As you can see, I am using the same logic we used to create the subnets earlier in this chapter using the `cidrsubnet` function in a `for_each` loop across the `subnets` variable.

So why would you want to do this?

As we saw when we used Terraform to deploy our WordPress workload in Microsoft Azure in *Chapter 4, Deploying to Microsoft Azure*, we had to build in logic to handle changes in subnet settings – in our case, this was to delegate a subnet for use with the Azure Database for MySQL – Flexible Server service.

The module provided by Claranet has this logic built in; for example, the code to add this would look like the following:

```
module "azure_network_subnet_001" {
  for_each            = var.subnets
  source              = "claranet/subnet/azurerm"
  environment         = var.environment
  location_short      = module.azure_region.location_short
  custom_subnet_name  = each.value.name
  client_name         = var.name
  stack               = var.project_name
  resource_group_name = module.rg.resource_group_name
  virtual_network_name = module.azure_virtual_network.virtual_network_
name
  subnet_cidr_list    = ["10.0.0.0/27"]
  subnet_delegation = {
    flexibleServers = [
      {
        name    = "Microsoft.DBforMySQL/flexibleServers"
        actions = ["Microsoft.Network/virtualNetworks/subnets/join/
action"]
      }
    ]
  }
}
```

Claranet has over 80 other modules for Microsoft Azure and Amazon Web Services on Terraform registry, and they are not the only provider to have modules published there – both other providers and individuals have modules published there that are all free to use.

You can also publish your own modules on the Terraform registry or even host them on GitHub as either public or private repositories; the advantage of using modules and taking this approach is that it enables you very quickly to develop your IaC deployments with consistent reusable components.

So, what about Ansible?

As already mentioned, you can use roles, which are distributed via Ansible Galaxy – there are far fewer roles there compared to the modules available in Terraform – but you can publish your own or reuse them locally.

Pop quiz

Before we finish the chapter, let's have a quick pop quiz:

1.   What is the name of the function we use to work with CIDR ranges in Terraform?

2.   When passing in variables at runtime, which tools use the `--extra-vars` flag, and which one uses `-var`?

3.   What key can be used to loop through a list or map of variables in Terraform tasks or modules?

4.   When working with NFS, which of the two public clouds requires a mount target to be configured?

5.   Azure Database for MySQL – Flexible Server requires us to do what to a subnet?

You can find the answers after the summary.

Summary

With the use of variables, modules, or roles, you can quickly build up your IaC deployments in a consistent way that can be shared with the rest of your team, allowing everyone to build their environments using a set of shared building blocks.

Another advantage of this approach is that you are deploying the same sort of infrastructure repeatedly for your project because you have multiple environments or multiple customers.

Having a set of variables per deployment changing things such as the **stock keeping units** (**SKUs**) or resource names, with everything else being the same, will save time and allow you to manage all your deployments centrally. We will look at how to centrally manage our deployment in our next chapter, *Chapter 7, Leveraging CI/CD in the Cloud*.

Before we move on, let us quickly summarize what we have discussed in this chapter. We started by clearing up what we mean by cloud-agnostic tools before looking at the difference between our Amazon Web Services and Microsoft Azure deployments.

We then discussed the differences in the approaches we need to take when choosing to use Terraform or Ansible; we also did a bit of a deep dive into how we can combine the two tools and use Ansible to manage our Terraform deployment.

Further reading

You can find more details on the services and documentation we have mentioned in this chapter at the following URLs:

- Terraform Registry: `https://registry.terraform.io`
- Claranet Terraform modules and providers: `https://registry.terraform.io/namespaces/claranet`
- Ansible Galaxy: `https://galaxy.ansible.com`
- Ansible Terraform module: `https://docs.ansible.com/ansible/latest/collections/community/general/terraform_module.html`
- ChatGPT: `https://openai.com/blog/chatgpt/`

Answers

Here are the answers to the pop quiz:

1. What is the name of the function we use to work with CIDR ranges in Terraform? The answer is `cidrsubnet`.

2. When passing in variables at runtime, which tools use the `--extra-vars` flag, and which one uses `-var`? The `--extra-vars` flag is used by Ansible, and `-var` is Terraform.

3. What key can be used to loop through a list or map of variables in Terraform tasks or modules? The key is `for_each` with the value being the variable you wish to loop through.

4. When working with NFS, which of the two public clouds requires a mount target to be configured? The answer is Amazon Web Services.

5. Azure Database for MySQL – Flexible Server requires us to do what to a subnet? Azure Database for MySQL – Flexible Server must have an entire subnet delegated to it using the `delegate` key.

Part 3:
CI/CD and Best Practices

In this part, we will look at using **Continuous Integration/Continuous Deployment (CI/CD)** hosted in the cloud. We will be using GitHub Actions to execute our Terraform and Ansible deployments.

We will then move on to discussing best practices and some common troubleshooting tips, before finally reviewing a few alternatives to Terraform and Ansible.

This part has the following chapters:

- *Chapter 7, Leveraging CI/CD in the Cloud*
- *Chapter 8, Common Troubleshooting Tips and Best Practices*
- *Chapter 9, Exploring Alternative Infrastructure-as-Code Tools*

7

Leveraging CI/CD in the Cloud

We have discovered that **Infrastructure as Code (IaC)** has become an essential practice for modern development, enabling developers to manage infrastructure through code rather than manually configuring it.

However, deploying our infrastructure from a local machine, which we have been doing until now, is no longer sufficient for large-scale systems.

This is where **Continuous Integration/Continuous Deployment (CI/CD)** comes into play; it automates the deployment process and provides consistent and reliable infrastructure deployment.

This chapter will explore leveraging CI/CD in the cloud to deploy our IaC. We will focus on the popular CI/CD tool, **GitHub Actions**, which can run workflows triggered by different events, such as pull requests or code commits. We will explore how to use GitHub Actions to run Terraform and Ansible code in both public clouds we covered in *Chapter 4, Deploying to Microsoft Azure*, and *Chapter 5, Deploying to Amazon Web Services*.

We will also cover essential security practices such as managing secrets in GitHub Actions and monitoring and maintaining the deployment once it runs. By the end of this chapter, you will understand how to leverage CI/CD in the cloud for your IaC projects using GitHub Actions.

We will cover the following topics in this chapter:

- Introducing GitHub Actions
- Running Terraform using GitHub Actions
- Running Ansible using GitHub Actions
- Security best practices

Before we roll our sleeves up and dive into the code, we should discuss the CI/CD tool we will use to deploy our infrastructure.

Technical requirements

The source code for this chapter is available here: `https://github.com/PacktPublishing/` `Infrastructure-as-Code-for-Beginners/tree/main/Chapter07`

Introducing GitHub Actions

So, what is GitHub Actions? GitHub Actions is an automation platform that allows developers to create workflows to automate software development tasks, which in our case, means managing and deploying our IaC workloads.

The beta of GitHub Actions was first launched in mid-2019. The initial release of GitHub Actions allowed a select number of developers to create and share actions that could be used to automate repetitive tasks in their development workflow. It was launched as a competitor to other popular automation platforms such as Jenkins, Travis CI, and CircleCI.

GitHub Actions is based on several concepts, and the ones which will be covered in detail in this chapter are the following:

- **Workflows**: These are a series of tasks that are automated using GitHub Actions. Workflows are defined in YAML files that are stored in the repository. Workflows can be triggered by various events, such as pushing code to the repository, creating a pull request, or scheduling a job.

- **Jobs**: These are the individual units of work performed within a workflow. A workflow can have multiple jobs, each running on a different platform or environment. Jobs can be run in parallel or sequentially, depending on the workflow requirements.

- **Steps**: These are the individual tasks that make up a job. Each step can be a shell command, a script, or an action. Actions are pre-built units of work that can be used to automate everyday development tasks, such as building and testing code, deploying applications, and sending notifications.

- **Events**: The trigger workflows and GitHub Actions support many event types, including pushes to the repository, pull requests, scheduled events, and manual triggers.

GitHub Actions is a powerful automation platform that allows developers to automate many tasks in their development workflow. With its flexible and customizable workflows, support for various events, and pre-built actions, GitHub Actions has become an essential tool for many teams.

With its continued development and new features, GitHub Actions is set to establish itself as the leading CI/CD automation platform.

Rather than talking any more about GitHub Actions, let's roll up our sleeves and look at how to run Terraform using it.

Running Terraform using GitHub Actions

In the last four chapters, we have talked about Terraform a lot – however, we are yet to address the elephant in the room – state files.

As we have been running Terraform locally, we haven't really needed to talk about state files in too much detail yet, so let's look at them now before we discuss how we can run Terraform using GitHub Actions.

Terraform state files

Every time we run Terraform, a file called `terraform.tfstate` is either created, updated, or read. It is a JSON formatted file containing information about the resources created or modified by Terraform. It includes details such as the IDs, IP addresses, and other metadata associated with each resource we manage with Terraform.

Terraform uses this file to keep track of the current state of the infrastructure to determine what changes must be made when you modify your infrastructure code.

The state file is absolutely critical to the correct operation of Terraform. It ensures that Terraform can accurately determine what changes need to be made to the infrastructure when you run the `terraform apply` command.

Without a state file, Terraform would be unable to determine what changes to make to your infrastructure, which could result in errors or unexpected behaviors – for example, the termination and redeployment of a resource.

It is also important to note that a Terraform state file should be treated as sensitive information; it contains details about your infrastructure resources and also potentially sensitive information, such as passwords if you have used Terraform to generate them.

This means we must ensure that the state file is securely stored and only accessible to authorized users.

So why are we only talking about this now?

Well, services such as GitHub Actions are designed to provide compute resources for a short amount of time to execute a workflow, making them ephemeral, which means that there is no fixed underlying storage, so once the workflow has been completed, the compute resource is terminated, and everything is lost.

To support this, Terraform allows you to use backends to store your state files; as you may have already guessed, the default storage option is local storage, which will store the file in the same folder as the Terraform code you are executing. You can also use external blob storage such as Amazon **Simple Storage Service (S3)** (`s3`) or Azure storage accounts (`azurerm`).

The following example shows how you would use an Azure storage account called `satfbeiac1234` in a resource group called `rg-terraform-state-uks`:

```
terraform {
  backend "azurerm" {
    resource_group_name   = "rg-terraform-state-uks"
    storage_account_name  = "satfbeiac1234"
    container_name        = "tfstate"
    key                   = "prod.terraform.tfstate"
  }
}
```

The `container_name` parameter, in the case of an Azure storage account, is the blob container, which, if you were thinking of the Azure storage account as a filesystem, then this would be the folder name, and `key` is the name of the file.

The configuration for Amazon S3 is not too dissimilar, as you can see from the following example:

```
terraform {
  backend "s3" {
    bucket = "tfbeiac1234"
    key    = "tfstate/prod.terraform.tfstate"
    region = "us-east-1"
  }
}
```

Here we are letting Terraform know the bucket name, the path to our file, and the region in which the Amazon S3 bucket is hosted.

One thing that needs to be added to the preceding code is how the Azure storage account or Amazon S3 bucket is created in the first place and how Terraform authenticates against the cloud provider to be able to read and write to the backend.

Rather than discuss here, let's dive into an example GitHub action and find out.

GitHub Actions

I am going to concentrate on Microsoft Azure in this chapter. So, as we are not using our locally installed copy of the Azure **command-line interface** (**CLI**), we need to generate some credentials to use and grant access to our Azure subscription.

> **Info**
> Please note that the **Universally Unique Identifiers** (**UUIDs**) in the following list are just examples; please ensure you replace them with your own where prompted.

To do this, we are going to create a service principal using the following command. When you run it, make sure you replace the subscription ID with your own subscription ID, which you can find in the Azure portal under **Subscriptions**:

```
$ az ad sp create-for-rbac --scopes /subscriptions/3a52ef17-7e42-4f89-
9a43-9a23c517cf1a --role Contributor
```

This will give something similar to the following output, which starts with an important message:

```
Creating 'Contributor' role assignment under scope '/
subscriptions/3a52ef17-7e42-4f89-9a43-9a23c517cf1a
The output includes credentials that you must protect. Be sure
that you do not include these credentials in your code or check
the credentials into your source control. For more information, see
https://aka.ms/azadsp-cli
```

As per the preceding message, ensure that you make a note of the output in a safe place, as it will be the only time you will be able to get the password that is generated:

```
{
    "appId": "019f16d2-552b-43ff-8eb8-6c87b13d47f9",
    "displayName": "azure-cli-2023-03-18-14-28-04",
    "password": "6t3Rq~vT.cL9y7zN_apCvGANvAg7_v6wiBb1eboQ",
    "tenant": "8a7e32c4-5732-4e57-8d8c-dfca4b1e4d4a"
}
```

Now that we have the details and have granted the newly created `Contributor` service principal access to our Azure subscription, we can move on to GitHub.

We first need to enter some secrets and variables in the GitHub repository to configure the GitHub action.

I have started with an empty GitHub repository called `Terraform-github-actions-example`; if you are following along, I recommend creating a test repo and copying the code from the repository accompanying this title across to your repo.

As mentioned, the first thing we need to do is add the secrets and variables. To do this, go to your repo and click on **Settings**. Once the **Settings** page is open, you should see **Secrets and variables** in the left-hand side menu; when you click on it, it will expand a submenu with **Actions**, **Codespaces**, and **Dependabot** listed.

As you might have guessed, you need to click on **Actions**. This should present you with something that looks like the following:

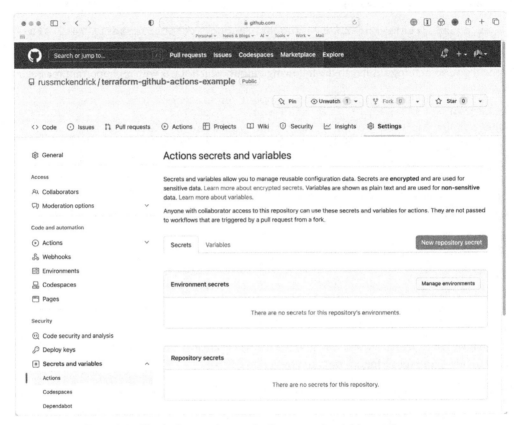

Figure 7.1 – The Actions option on the Secrets and variables settings page

If you click on the **New repository secret** button and enter the secrets detailed in the following table, make sure that you enter the name exactly like it is written in the table, as the GitHub action workflow code will reference the name when it is executed:

Name	Secret Content
ARM_CLIENT_ID	This is the `appId` value from the output of the command that added the service principal. From the example output, this would be `019f16d2-552b-43ff-8eb8-6c87b13d47f9`.

ARM_CLIENT_SECRET	This is the password from the output of when we ran the command that added the service principal. From the example output, this would be 6t3Rq~vT.cL9y7zN_apCvGANvAg7_v6wiBb1eboQ.
ARM_SUBSCRIPTION_ID	This is the subscription ID you used as the scope to add the service principal. From the example output, this would be 3a52ef17-7e42-4f89-9a43-9a23c517cf1a.
ARM_TENANT_ID	This is the tenant from the output of when we ran the command that added the service principal. From the example output, this would be 8a7e32c4-5732-4e57-8d8c-dfca4b1e4d4a.

Once you have entered the four secrets detailed in the preceding table, you will use the credentials to authenticate against your Azure account and make changes. We can now enter the variables; these detail the storage account and don't need to be stored as secrets:

Name	Value Content
BACKEND_AZURE_RESOURCE_GROUP_NAME	This is the name of the resource group that will be created to host the storage account we will use for the Terraform state file, for example; rg-terraform-state-uks.
BACKEND_AZURE_LOCATION	The region the resources are going to be launched in. For example, uksouth.
BACKEND_AZURE_STORAGE_ACCOUNT	The name of the storage account you create must be unique across all of Azure; otherwise, you will get an error. For example, satfstate180323.
BACKEND_AZURE_CONTAINER_NAME	The name of the container where the file will be stored, for example, tfstate.
BACKEND_AZURE_STATE_FILE_NAME	The name of Terraform state file itself, for example, ghact.tfstate.

So now that we have everything that we need secret- and variable-wise in the GitHub repository, we can look at the workflow itself.

GitHub action workflows are **YAML Ain't Markup Language** or **Yet Another Markup Language** (**YAML**) files (depending on the explanation you read).

> **Information**
>
> YAML is a human-readable data serialization format that uses indentation to convey structure, popular for configuration files, data exchange, and applications requiring simple data representation.

First, we have some basic configuration; here, we use name to name the workflow and define on to specify on what action the workflow should trigger:

```
name: "Terraform Plan/Apply"
on:
  push:
    branches:
      - main
  pull_request:
    branches:
      - main
```

As you can see, the workflow will be triggered as defined in the on section of the YAML file, on a push or pull_request to the main branch. Now that we have defined when the workflow will be triggered, we can now define the three jobs that go to make up the workflow, starting with the job that checks the presence of the storage account we are going to use for the backend Terraform state file – and it if it is not there, create one.

> **Please note**
>
> Indentation is really important when it comes to the structure of the YAML file; however, while working through the structure across the following pages, I will be removing some of the indentations to make it more readable – please refer to the code in the GitHub repository that accompanies this title for the correct format and indentation.

First, we define the jobs and some basic configurations:

```
jobs:
  check_storage_account:
    name: "Check for Azure storage account"
    runs-on: ubuntu-latest
    defaults:
      run:
        shell: bash
```

Here we are giving the job an internal reference, check_storage_account, and telling it to run on the latest version of Ubuntu and use bash as the default shell.

The check_storage_account job is made up of two steps, step one being as follows:

```
steps:
    - name: Login to Azure using a service principal
      uses: Azure/login@v1
      with:
        creds: '{"clientId":"${{ secrets.ARM_CLIENT_
ID }}","clientSecret":"${{ secrets.ARM_CLIENT_SECRET
}}","subscriptionId":"${{ secrets.ARM_SUBSCRIPTION_ID
}}","tenantId":"${{ secrets.ARM_TENANT_ID }}"}'
```

Here we are using the Azure/login@v1 task to log in to our Azure account using the secrets we defined in GitHub. These are referred to by using ${{ secrets.ARM_CLIENT_ID }}. The next step uses the variables we added to GitHub and the Azure/CLI@v1 task to check for the presence of the storage account.

If it is not there, it will be created, and the resources already exist, then the task will progress to the next step:

```
- name: Create Azure storage account
  uses: Azure/CLI@v1
  with:
    inlineScript: |
      az group create --name ${{ vars.BACKEND_AZURE_RESOURCE_GROUP_
NAME }} --location ${{ vars.BACKEND_AZURE_LOCATION }}
      az storage account create --name ${{ vars.BACKEND_AZURE_STORAGE_
ACCOUNT }} --resource-group ${{ vars.BACKEND_AZURE_RESOURCE_GROUP_NAME
}} --location ${{ vars.BACKEND_AZURE_LOCATION }} --sku Standard_LRS
      az storage container create --name ${{ vars.BACKEND_AZURE_
CONTAINER_NAME }} --account-name ${{ vars.BACKEND_AZURE_STORAGE_
ACCOUNT }}
```

As this step is running on the same Ubuntu instance as the first step, which logged into Azure, we don't need to authenticate again – instead, we can just run the Azure CLI commands we need:

1. Create or check for the presence of the resource group to host our storage account using az group create.

2. Create or check for the presence of the storage account using az storage account create.

3. Create or check for the presence of the container in the storage account using az storage container create.

Now that we know we have the storage account in place for the Terraform backend state file storage, we can proceed with the next job, which runs the terraform_plan command and stores the output with the workflow run.

As per the last job, we need to set up some basic configurations, such as the job name and reference, what operating system to use, and also some additional bits:

```
terraform_plan:
  name: "Terraform Plan"
  needs: check_storage_account
  runs-on: ubuntu-latest
  env:
    ARM_CLIENT_ID: "${{ secrets.ARM_CLIENT_ID }}"
    ARM_CLIENT_SECRET: "${{ secrets.ARM_CLIENT_SECRET }}"
    ARM_SUBSCRIPTION_ID: "${{ secrets.ARM_SUBSCRIPTION_ID }}"
    ARM_TENANT_ID: "${{ secrets.ARM_TENANT_ID }}"
  defaults:
    run:
      shell: bash
```

As you can see, we are setting some environment variables containing the credentials needed to log in to Azure; why are we doing that again when we already authenticated during the last job?

The reason is that once the last task in the last job finished, the compute resource running the job was terminated, and when this job started, a new resource was spun up, meaning that everything from the last job was lost.

Now that we have defined the basics for the `terraform_plan` job, we can work through the steps:

```
steps:
  - name: Checkout the code
    id: checkout
    uses: actions/checkout@v3
```

This simple step checks out the repository from which we are running the action; the repository contains the Terraform we will execute during the workflow.

Now we have the code we need to install Terraform. To do this, we use the `hashicorp/setup-terraform@v2` task:

```
- name: Setup Terraform
  id: setup
  uses: hashicorp/setup-terraform@v2
  with:
    terraform_wrapper: false
```

So far, so good; as per when we were running Terraform on our local machine, we now need to run the `terraform init` command:

```
- name: Terraform Init
  id: init
```

```
    run: terraform init -backend-config="resource_group_name=${{ vars.
BACKEND_AZURE_RESOURCE_GROUP_NAME }}" -backend-config="storage_
account_name=${{ vars.BACKEND_AZURE_STORAGE_ACCOUNT }}" -backend-
config="container_name=${{ vars.BACKEND_AZURE_CONTAINER_NAME }}"
    -backend-config="key=${{ vars.BACKEND_AZURE_STATE_FILE_NAME }}"
```

As you can see, we have appended quite a bit to the end of the terraform init command – this sets up our backend using the variables we defined in GitHub for the duration of the job, meaning Terraform will use the remote backend and not the local one.

Next up, we need to run the terraform plan command to figure out what needs to happen during the workflow execution:

```
- name: Terraform Plan
  id: tf-plan
  run: |
    export exitcode=0
    terraform plan -detailed-exitcode -no-color -out tfplan || export
exitcode=$?
    echo "exitcode=$exitcode" >> $GITHUB_OUTPUT
```

You will have noticed that we are wrapping a little logic around the command to figure out the exit code. We need to do this because we need to know whether we should stop the execution of the workflow if there is an error, which is what the final piece of the code in the step does:

```
    if [ $exitcode -eq 1 ]; then
      echo Terraform Plan Failed!
      exit 1
    else
      exit 0
    fi
```

So now we know whether there are any obvious errors or everything is OK, and we have a copy of the Terraform plan file; what's next?

As we have already mentioned, when we run the next job, we will be starting from scratch, and as we need a copy of the Terraform plan file, we should copy it from the compute resource:

```
- name: Publish Terraform Plan
  uses: actions/upload-artifact@v3
  with:
    name: tfplan
    path: tfplan
```

We are using the actions/upload-artifact@v3 task to copy the file called tfplan to the workflow execution as an artifact; in subsequent tasks and jobs, we can download the file and use it without committing to the code repo itself.

The next task, at first glance, may seem a little redundant:

```
- name: Create String Output
  id: tf-plan-string
  run: |
    TERRAFORM_PLAN=$(terraform show -no-color tfplan)
    delimiter="$(openssl rand -hex 8)"
    echo "summary<<${delimiter}" >> $GITHUB_OUTPUT
    echo "## Terraform Plan Output" >> $GITHUB_OUTPUT
    echo "<details><summary>Click to expand</summary>" >> $GITHUB_
OUTPUT
    echo "" >> $GITHUB_OUTPUT
    echo '```terraform' >> $GITHUB_OUTPUT
    echo "$TERRAFORM_PLAN" >> $GITHUB_OUTPUT
    echo '```' >> $GITHUB_OUTPUT
    echo "</details>" >> $GITHUB_OUTPUT
    echo "${delimiter}" >> $GITHUB_OUTPUT
```

The task appears to be doing something with the Terraform plan file, but what?

One of the advantages of using a system such as GitHub actions is that you can publish artifacts and also publish other outputs – in this case, we are taking the list of changes logged within the plan file and formatting it for use as a workflow summary.

The next and final task within this job is to take the summary we have just generated and publish it back to GitHub:

```
- name: Publish Terraform Plan to Task Summary
  env:
    SUMMARY: ${{ steps.tf-plan-string.outputs.summary }}
  run: |
    echo "$SUMMARY" >> $GITHUB_STEP_SUMMARY
```

Now, all we have left is to run the `terraform apply` command – this is the last job of our workflow, and it shares many of the steps with the previous job.

However, there are some changes to the configuration that we should highlight:

```
terraform-apply:
  name: "Terraform Apply"
  if: github.ref == 'refs/heads/main'
  runs-on: ubuntu-latest
  env:
    ARM_CLIENT_ID: "${{ secrets.ARM_CLIENT_ID }}"
    ARM_CLIENT_SECRET: "${{ secrets.ARM_CLIENT_SECRET }}"
    ARM_SUBSCRIPTION_ID: "${{ secrets.ARM_SUBSCRIPTION_ID }}"
    ARM_TENANT_ID: "${{ secrets.ARM_TENANT_ID }}"
```

```
    needs: [terraform_plan]
```

As you can see, we have added the `if` and `needs` statements. The `if` statement verifies that we are 100% working with the correct branch, and the `needs` statement ensures that the `terraform_plan` job has been successfully executed, meaning that we will have the Terraform plan file.

The first three steps are ones that we have already covered, them being the following:

1. Check out the code
2. Set up Terraform
3. `terraform init`

Next, we need to download the Terraform plan file:

```
- name: Download Terraform Plan
  uses: actions/download-artifact@v3
  with:
    name: tfplan
```

With the plan file downloaded, we can now perform the final task of the workflow, which is to run the `terraform apply` command and deploy the changes, if any, detailed within the plan file.

Given the number of tasks it has taken for us to get to this point, the final task is quite simple:

```
- name: Terraform Apply
  run: terraform apply -auto-approve tfplan
```

As you can see, we run `terraform apply` with the `-auto-approve` flag; if we don't, then Terraform will quite happily sit there for an hour waiting for someone to type `Yes`, which will never happen as this is not an interactive terminal.

We are then telling it to load in the file called `tfplan`, which means that we do not need to run the `terraform plan` command for a second time as we already know what will change/be applied during the execution.

So, what changes to our Terraform code are needed for this to work?

Just the one we need to tweak our code to use the `azurerm` backend; this makes the top of our `main.tf` file look like the following:

```
terraform {
  required_version = ">=1.0"
  required_providers {
    azurerm = {
      source  = "hashicorp/azurerm"
      version = "~>3.0"
    }
```

```
    }
  backend "azurerm" {}
}
```

The rest of the code remains as is. We then need to take a workflow YAML file and place it in a folder called `.github/workflows` at the top level of our repository. I have named the file `action.yml`.

Please note

In the repo accompanying this title, the folder name purposely has the "`.`" removed, so the GitHub action is not registered. When you copy to your repo, please ensure you rename the `github` folder to `.github`; otherwise, the action won't be registered, and the workflow will not run.

So let's run it the first time you check in the `action.yml` file. It will create the action and run – this can be confirmed by the dot next to the commit ID, which in the following example screen is referenced as **b4900e8**:

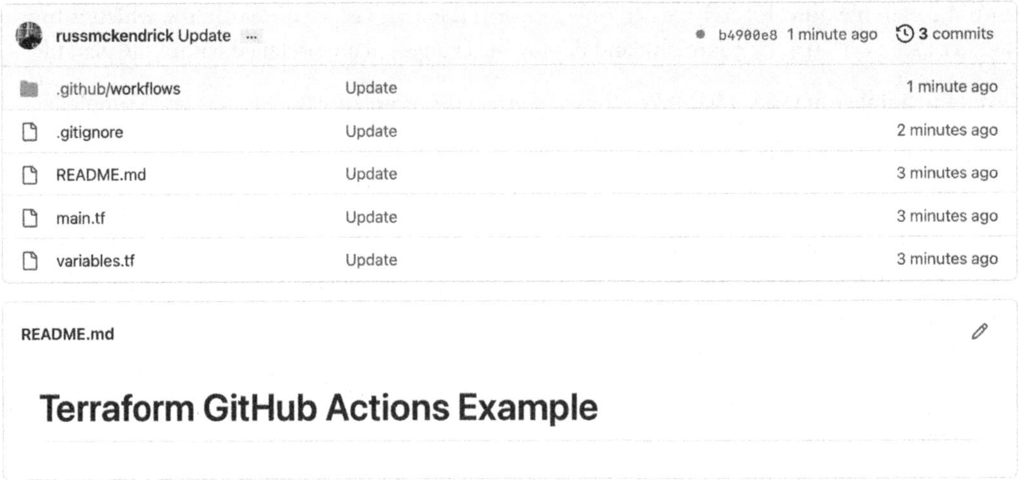

russmckendrick Update ···		● b4900e8 1 minute ago ⟳ 3 commits	
.github/workflows	Update		1 minute ago
.gitignore	Update		2 minutes ago
README.md	Update		3 minutes ago
main.tf	Update		3 minutes ago
variables.tf	Update		3 minutes ago

README.md

Terraform GitHub Actions Example

Figure 7.2 – Checking in and running the workflow

If everything has run as expected, clicking the **Actions** tab at the top of the repo page should show you something like the following:

Figure 7.3 – Viewing the workflow runs

Clicking on the workflow run will take you to the execution **Summary** page; for me, this looked like the following:

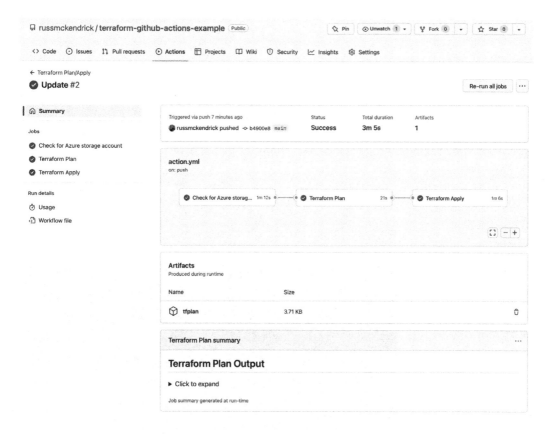

Figure 7.4 – Reviewing the workflow execution

As you can see, we have the three jobs listed and the artifacts and the custom summary we published from the **Terraform Plan** job. Clicking **Click to expand** will show you the entire output of the `terraform plan` command:

Figure 7.5 – Viewing the output of Terraform Plan

Also, if you click on any of the job names, it will show the output of each of the tasks:

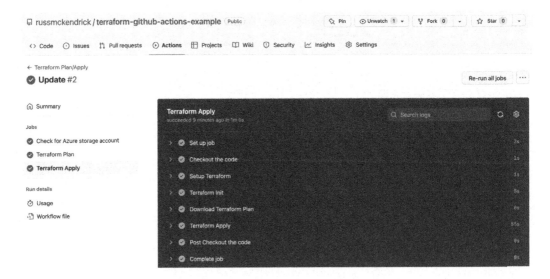

Figure 7.6 – Viewing the output of Terraform Plan

I recommend clicking around and reviewing precisely what the GitHub action workflow executed, as it gives pretty detailed information.

Finally, if you check the Azure portal, you should see the resource group, storage account, and container where there should be a single file called `ghact.tfstate`:

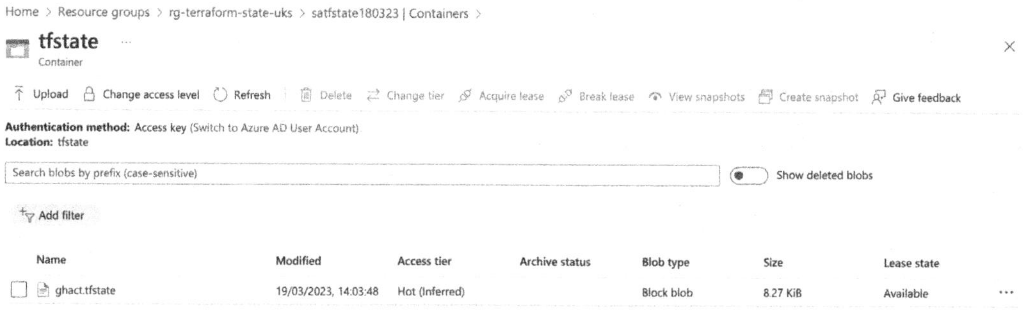

Figure 7.7 – Checking the Terraform state file in the Azure portal

That concludes using GitHub actions to run Terraform; before we finish the chapter, let's look at a workflow that runs Ansible.

Running Ansible using GitHub Actions

Ansible doesn't have a concept of state files, so this will simplify our GitHub action workflow. As we are using Microsoft Azure again, you must set up the `ARM_CLIENT_ID`, `ARM_CLIENT_SECRET`, `ARM_SUBSCRIPTION_ID`, and `ARM_TENANT_ID` secrets in your GitHub repository as we did in the last section before progressing.

Once they are there, we can move on to the workflow itself; as with the Terraform workflow, we start by setting some basic configurations:

```
name: "Ansible Playbook Run"
on:
  push:
    branches:
      - main
  pull_request:
    branches:
      - main
```

Then we define the job; that's right, there is only one job for this workflow:

```
jobs:
  run_ansible_playbook:
    name: "Run Ansible Playbook"
    runs-on: ubuntu-latest
    defaults:
      run:
        shell: bash
```

So far, not much is different, so let's move on to the steps. First, we check out the code:

```
steps:
  - name: Checkout the code
    id: checkout
    uses: actions/checkout@v3
```

Here we hit our first difference; as Ansible is written in Python, we need to make sure that Python is installed and reasonably up to date. For this, we will use the `actions/setup-python@v4` task:

```
- name: Ensure that Python 3.10 is installed
  uses: actions/setup-python@v4
  with:
    python-version: "3.10"
```

The next step is to log in to Azure, this is an exact copy of the *Log in to Azure using a service principal* step from the Terraform workflow in the previous section of this chapter, so I will not repeat the code here.

Next, we need to install Ansible itself – we are doing this using the `pip` command; the step looks like the following:

```
- name: Install Ansible
  run: pip install ansible
```

Once Ansible is installed, we can then run the `ansible-galaxy` command to install the Azure Collection – this step is not too different from when installed it locally:

```
- name: Install Azure Collection
  run: ansible-galaxy collection install azure.azcollection
```

As you may have guessed, once the Azure Collection is installed, we need to install the Python modules needed for the collection to function:

```
- name: Install Azure Requirements
  run: pip install -r ~/.ansible/collections/ansible_collections/
azure/azcollection/requirements-azure.txt
```

Once everything we need to run the playbook is installed, we can run the task; this step looks like the *Create String Output* step in the Terraform workflow, as we want to capture the output of the `ansible-playbook` command and store it within the workflow summary:

```
- name: Run the playbook (with ansible-playbook)
  id: ansible-playbook-run
  run: |
    ANSIBLE_OUTPUT=$(ansible-playbook site.yml)
    delimiter="$(openssl rand -hex 8)"
    echo "summary<<${delimiter}" >> $GITHUB_OUTPUT
    echo "## Ansible Playbook Output" >> $GITHUB_OUTPUT
    echo "<details><summary>Click to expand</summary>" >> $GITHUB_OUTPUT
    echo "" >> $GITHUB_OUTPUT
    echo '```' >> $GITHUB_OUTPUT
    echo "$ANSIBLE_OUTPUT" >> $GITHUB_OUTPUT
    echo '```' >> $GITHUB_OUTPUT
    echo "</details>" >> $GITHUB_OUTPUT
    echo "${delimiter}" >> $GITHUB_OUTPUT
```

The final step in the workflow is to publish the summary:

```
- name: Publish Ansible Playbook run to Task Summary
  env:
    SUMMARY: ${{ steps.ansible-playbook-run.outputs.summary }}
  run: |
    echo "$SUMMARY" >> $GITHUB_STEP_SUMMARY
```

That is it; as you can see, the workflow has fewer jobs and steps as we don't have to take into consideration either the backend storage or publishing plan file as an artifact as we did for the Terraform workflow.

Running the workflow should give you something like the following output:

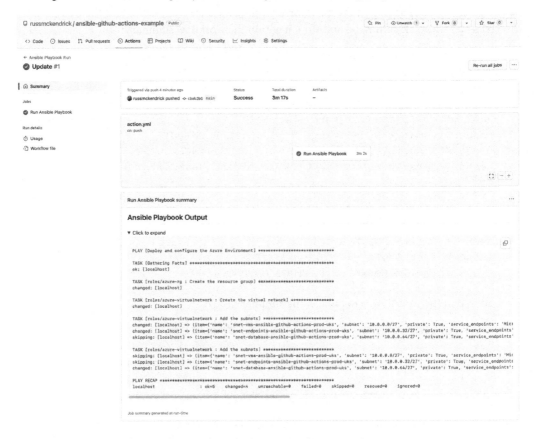

Figure 7.8 – Running the Ansible Playbook using GitHub Actions

Again, the folder name of the repo accompanying this title purposely has the . character removed from the start of the folder name, so the GitHub action is not registered. If you are following along in your repo, per the Terraform GitHub Action workflow, you must rename this folder to .github when committing to your repo to register the action.

Now that we have run our workflows using Terraform and Ansible, let's quickly discuss some best practices.

Security best practices

When we worked through the Terraform and Ansible workflows, we discussed adding repository secrets to our GitHub repository. All sensitive information should be stored within secrets outside of using an external source for your secrets, such as Azure Key Vault, AWS Secrets Manager, or HashiCorp Vault.

The advantage of this is that the secrets will remain hidden, but the code will also be able to consume them. Great, you may think to yourself.

But anyone who has been granted write access to the repo will also be able to consume them (though not view the contents), so please be careful when granting access to your IaC CI/CD pipelines as they will have a high level of access to your cloud resource via your workflows, so please ensure that you only grant access to trusted members of your teams.

Pop quiz

Before we finish the chapter, let's have a quick pop quiz:

1. When writing YAML, what is it essential to keep an eye on?

2. When it comes to credentials, what should you never do?

3. What is the folder name in which the GitHub action should be stored?

Summary

While we spent a lot of the initial part of the chapter discussing how Terraform works, once we got onto working through the GitHub Actions workflows, I am sure that you started to see the benefits of running our IaC from a centrally accessible location rather than your local machine.

Once we discussed Terraform's requirements, we configured repository secrets and variables in GitHub. Then we worked through the various jobs and steps to make up the workflow that manages the storage account, where we stored the Terraform state and executed the Terraform deployment.

We then took everything we learned and covered in Terraform and applied it to Ansible before finally discussing a vital security point – be careful what access you give to your IaC GitHub actions!

There are some points we needed more time to cover, such as monitoring; for example, it is relatively straightforward to hook your GitHub Actions into messaging services such as Microsoft Teams or Slack to get real-time feedback on workflow runs – there are links to the GitHub Actions Marketplace tasks for Microsoft Teams and Slack in the further reading section if you want to have a go at hooking your workflows into your preferred messaging service.

This is not only a great way of extending your IaC deployments to other team members, but it also works as a system that tracks changes as the workflow runs, which contains a summary of each execution that will be stored for a while.

In the next and penultimate chapter, we will look at common troubleshooting tips and tricks.

Further reading

You can find more details on the tasks we have used in the steps in this chapter at the following URLs:

- `https://github.com/marketplace/`
- `https://github.com/marketplace/actions/azure-login`
- `https://github.com/marketplace/actions/azure-cli-action`
- `https://github.com/marketplace/actions/checkout`
- `https://github.com/marketplace/actions/upload-a-build-artifact`
- `https://github.com/marketplace/actions/download-a-build-artifact`
- `https://github.com/marketplace/actions/setup-python`
- `https://github.com/marketplace/actions/microsoft-teams-deploy-card`
- `https://github.com/marketplace/actions/slack-notify`

Answers

Here are the answers to the pop quiz:

1. When writing YAML, what is it essential to keep an eye on? Indentation! The structure of your YAML file is critical – if you get it wrong, even by a single character, you will get errors.

2. When it comes to credentials, what should you never do? Embed them into your code! You need to use an external secret management system.

3. What is the folder name in which the GitHub action should be stored? Your YAML files should be stored in the `.github/workflows` folder.

8

Common Troubleshooting Tips and Best Practices

So far in the book, we have primarily discussed code samples that have been pre-written and shared through the GitHub repository accompanying this book. As you progress in your journey with **Infrastructure as Code (IaC)**, it's essential to understand that writing and planning your IaC projects will involve a learning curve and some inevitable debugging.

In this chapter, the second-to-last one, we'll look at some essential aspects to help you better plan, write, and troubleshoot your IaC projects.

We'll cover three key areas to ensure you're well equipped to handle any challenges that may arise during the process:

- Infrastructure as Code – best practices and troubleshooting

- Terraform – best practices and troubleshooting

- Ansible – best practices and troubleshooting

Throughout this chapter, you'll notice some common themes and advice that apply to both Terraform and Ansible, as they are both IaC tools. However, each tool uniquely interacts with your resources, resulting in some differences in the approaches and techniques you'll use when troubleshooting.

By the end of this chapter, you'll be well prepared to tackle the challenges of implementing IaC projects using these powerful tools.

Technical requirements

The source code for this chapter is available here:

```
https://github.com/PacktPublishing/Infrastructure-as-Code-for-
Beginners/tree/main/Chapter08
```

Infrastructure as Code – best practices and troubleshooting

Let us start by discussing some general IaC best practices that can apply to various tools and platforms.

General IaC best practices

There are some common threads that we have already touched upon here, but it is essential to bring them up again as they are important:

- **Version control**: Make sure you use a version control system such as Git or one of the other available systems such as Mercurial, Subversion, or Azure DevOps Server, which was previously known as **Team Foundation Server (TFS)**, to name a few of the more common ones, to store and manage your infrastructure code.

  The odds of you, either personally or within the business, already using version control for your other projects is extremely high if you are taking steps to both define and deploy your infrastructure in and as code. This means that you have experience with version control and access to the tools, processes, and procedures to maintain code using version control.

  Employing version control enables collaboration, change tracking, and easy rollback to previous versions if needed.

- **Documentation**: You can approach documentation in several ways, and it doesn't matter how you do it, just as long as you do it!

  My personal approach to documenting my IaC deployments is to try and keep as much of the documentation within the code as possible using both in-comments and making sure that sections, tasks, functions, or variables are as clearly named and descriptive as possible while keeping within any of the constraints of the tool I am using.

  Also, depending on the complexity, I will summarise what the code does, attach it as a README file, and commit it to version control.

  The reason I do this is that while it is easy to keep track of what is going on while you are working on the project, when it comes to someone else picking it up – or even you revisiting it yourself after a few months of being away from the project – it can sometimes take them a little time to get their bearings.

  Your approach may differ, which leads nicely into the next piece of best practice.

- **Code reviews**: I recommend conducting regular code reviews to maintain code quality, ensure compliance with best practices, and share knowledge among team members.

You may already have processes to enforce this across other types of development within the business, such as your applications. It is just as crucial that the same principles govern your IaC projects as you may be asked to demonstrate that your code adheres to any guidelines that your application has to follow for compliance reasons. After all, your IaC project will be deploying and maintaining the resources your application will run upon.

- **Modularity**: When you write your infrastructure code, do it in smaller, reusable modules. This promotes reusability, maintainability, and better organization of your code base. We covered this in *Chapter 6, Building upon the Foundations*.

- **Continuous Integration and Continuous Deployment (CI/CD)**: We discussed this at length in the previous chapter, *Chapter 7, Leveraging CI/CD in the Cloud*. Even for development purposes, if you have your code in source control, you should ideally be leveraging CI/CD.

- **Testing**: In the perfect world, you should implement automated testing for your infrastructure code to validate its correctness, identify issues early, and increase its overall reliability. If you are using version control and CI/CD, you already have most of the tools to make this process easy. For example, in *Chapter 7, Leveraging CI/CD in the Cloud*, we had some break-points when running Terraform plan to catch potential issues.

- **Monitoring and logging**: Implement monitoring and logging solutions to track executions and detect issues allowing you to troubleshoot problems promptly. In *Chapter 7, Leveraging CI/CD in the Cloud*, our CI/CD pipelines kept a log of everything that happened during the execution. In Terraform's case, we generated and attached a snapshot of the plan file – this level of information can be very valuable when trying to figure out what would happen if something unexpected happened.

- **The principle of least privilege**: Limit access to resources and creation by granting the minimum necessary permissions for your infrastructure code executions to interact with the components they are working with.

  Depending on your target infrastructure, this may only sometimes be possible, but most cloud providers allow you to be very granular with the permissions. Also, depending on what you are deploying, this could take a little trial and error – but in the long run, it is worth investing the time in looking at it from a security point of view.

- **Immutable infrastructure**: Rather than updating existing infrastructure, create new infrastructure to replace the old one and reroute requests to it. This reduces the risk of errors due to configuration drift and forces deployments to be more predictable.

  This approach depends on your application, and it may only sometimes be practical for you to fully implement this approach. Still, the more your infrastructure components you can make immutable, the easier it will be for you to scale out and back down.

- **Secure by Design (SBD)**: As you write your infrastructure code, incorporate security best practices and tools from the beginning, such as encryption, identity management, and network segmentation if possible, and as already mentioned, focus on making these parts of your code as modular as possible so that you can easily reuse them across your projects.

Now that we have established some general best practices, let us move on and discuss some general troubleshooting tips.

General IaC troubleshooting tips

The following are some general troubleshooting tips, tricks, and approaches. As we are talking about general IaC tips, many of them are more preventive than tasks you would do to debug an issue:

- **Avoid hardcoding sensitive information**: Use secret management tools such as Azure Key Vault, HashiCorp Vault, or AWS Secrets Manager to securely store and retrieve sensitive information at runtime or use your infrastructure code to configure your resources to use the secret management tools directly.

  While it goes without saying that you shouldn't hardcode sensitive information such as passwords, private information, or secrets directly within your code (Ansible could be an exception, but more on that in the *Ansible – best practices and troubleshooting* section), there are advantages to using secret management tools – the biggest one is for things such as certificate management.

  Imagine it's a day or two before your SSL certificate expires, and you are rushing to get all resources that reference it updated. Using your target platform's secret store may mean that you only have to update the certificate; then all resources that use the certificate are automatically updated.

- **Keep dependencies up-to-date**: Throughout *Chapter 4, Deploying to Microsoft Azure*, and *Chapter 5, Deploying to Amazon Web Services*, you will have noticed that our infrastructure code utilized a lot of different tasks and modules.

  Regularly updating your dependencies will help you avoid security vulnerabilities and compatibility issues. As your target cloud APIs are updated, you may find that your code has issues or no longer works.

- **Don't overcomplicate your infrastructure code**: Keep your infrastructure code as simple as possible and avoid unnecessary complexity that may be difficult to maintain and troubleshoot should there be issues.

  It may look "cool" to build lots of logic or loops into your IaC. Still, it only takes a slight change as part of a tool or dependency update for it to come tumbling down – the more effort and time needed to code something, the more effort you are likely to put into debugging and refactoring it if and when there are problems.

Trust me, from experience, your future self will thank you for this.

- **Maintain a clean, well-organized code base**: Consistently use naming conventions, follow a directory structure, and remove obsolete code.

  Anyone within your team needs to be able to pick up your code and know what is going on without having seen it; you will not always be the only one who looks into any problems with your code.

  You want to avoid creating more work for whoever picks it up, as they will likely already be under pressure because someone has reported a problem.

- **Don't ignore error or warning messages**: Address any messages, especially non-breaking warning messages in your infrastructure code, promptly to prevent future issues.

  Most tools will stop execution when there are errors. However, most will also print warnings – these could be just be small things such as letting you know that functionality you are using will be deprecated or changed in future releases, and warnings will not stop execution. Still, they need to be addressed just like any errors you receive; it isn't every day you get the chance to avoid future errors, so take it.

  Finally, and this goes without saying, **communicate with your team**. Regularly communicate with your team about infrastructure changes, potential issues, and best practices to ensure everyone is on the same page when it comes to your IaC. You do not want to be a single point of failure, nor do you want to set your team up for failure should there be any problems.

Now that we have worked through the general best practices and troubleshooting tips, let us look at some of the things you should consider when using the two tools we have discussed in the book, starting with Terraform.

Terraform – best practices and troubleshooting

We will cover some of the recommendations we have already touched upon in the *General IaC best practices* section. Still, as mentioned at the start of the chapter, we will go into more detail about how they apply to just Terraform.

Terraform – best practices

Here are some best practices for approaching your Terraform deployment:

- **Use a modular approach**: Break down infrastructure into reusable modules, simplifying code maintenance and enabling reusability across different environments.

As we discussed in *Chapter 6, Building upon the Foundations*, Terraform modules can be hosted in the Terraform Registry or, which I have not mentioned, privately in your own Git repository. The following example code downloads the module from GitHub using **Secure Shell (SSH)**:

```
module "somefunction" {
  source = "git@github.com:someuser/tfmodule.git"
}
```

Assuming you are executing your Terraform code from somewhere that has access to the repository, it will download and use it.

This allows you to build a library of reusable modules for use across all of your projects, and it also allows you to share modules with the rest of your teams.

- **Keep a consistent naming convention**: Using a consistent naming convention for resources and modules improves readability and maintainability.

  Depending on the size of the team working on your infrastructure code, you should establish a style guide and guidelines for developing and maintaining your Terraform infrastructure code.

- **Manage state files securely**: We have already discussed storing your state files remotely in a backend such as an Azure storage account or AWS **Simple Storage Service (S3)** in *Chapter 7, Leveraging CI/CD in the Cloud*.

  Most of the supported backend services allow you to enable versioning and force encryption to ensure data integrity and security – make sure that this is enabled. Most services do it by default, but it is best to double-check.

  Also, there is another service that should have been mentioned: Terraform Cloud. HashiCorp (the makers of Terraform), has a cloud service that can store your state files securely and also act as a remote execution environment for your Terraform run. There are both free and paid options, and if you can use them, I recommend taking a look.

- **Plan, review, and test**: Before applying changes, ensure that you use the `terraform plan` command to visualize the potential impact of the code run. Use code reviews and automated testing to validate changes and minimize the risk of errors or something unexpected.

- **Use provider and resource version pinning**: While Terraform development rates differ from provider to provider, you may find that breaking changes are introduced.

  You should lock the versions of providers within your infrastructure code and define an explicit version number when registering the providers used to ensure a consistent and stable infrastructure.

- **Leverage built-in Terraform functions**: Use Terraform functions such as `lookup`, `count`, and `for_each` to reduce complexity and improve flexibility. There are also functions, as we discussed in *Chapter 4, Deploying to Microsoft Azure*, and *Chapter 5, Deploying to Amazon Web Services*, where you work out **Classless Inter-Domain Routing (CIDR)** ranges and perform transformations on input and output variables – all of which can help reduce the number of variables you have to define.

Terraform – troubleshooting

Here are some of the best practices for approaching your Terraform deployment:

- **Avoid hardcoding sensitive information**: As you may have already guessed, this is a common but significant recurring theme; *please do not do it!* Instead, with Terraform, you can use environment variables or secret management tools to avoid exposing sensitive data in your code.

- **Manage dependencies correctly**: Understand implicit and explicit dependencies within your deployment and use the `depends_on` parameter when necessary to avoid issues related to resource ordering.

  We discussed this in *Chapter 2, Ansible and Terraform beyond the Documentation*, in the *Fixing the error* section.

- **Be cautious with terraform destroy**: Accidental resource destruction can have severe consequences. Use safeguards such as `prevent_destroy` and ensure proper access controls.

  The following is an example of how you would use `prevent_destroy` to protect against the accidental deletion of an Azure storage account:

```
resource "azurerm_storage_account" "example" {
  name                     = "saiacforbeg2022111534"
  resource_group_name      = azurerm_resource_group.example.name
  location                 = azurerm_resource_group.example.
location
  account_tier             = "Standard"
  account_replication_type = "GRS"
  lifecycle {
    prevent_destroy = true
  }
}
```

  You would receive an error if you attempted to run `terraform destroy` against the resource, which is much better than unexpectedly deleting the resource.

  Please note that this is not a resource lock at the cloud-provider level; you are just instructing Terraform that it can't destroy the resource on execution.

- **Monitor resource limits**: Be aware of provider-specific limits and quotas, which could lead to resource provisioning failures if they are hit.

  Errors while provisioning resources due to limits or quotas could result in a corrupted state file, which may not be easily recoverable depending on the resource you are targeting.

- **Monitor drift**: Detect and address configuration drift by regularly running `terraform refresh` and `terraform plan`. You could do this using CI/CD and have it alert depending on the output.

- **Watch for state file conflicts**: If multiple team members work on the same infrastructure, use remote state backends with locking mechanisms to prevent conflicting changes. Most backends support this by default, but to avoid state file corruption for production resources, I recommend triple-checking.

- **If possible, avoid using multiple provisioning tools**: Mixing Terraform with other provisioning tools (for example, CloudFormation or **Azure Resource Manager** (**ARM**) templates) can cause conflicts and unexpected behavior on subsequent executions. Stick to one provisioning tool for consistency and predictability, and if possible, attempt to find a workaround for your reason to deploy using the multiple tools in this place. This is a slightly different use case than what we discussed in *Chapter 6, Building upon the Foundations*, where we used Ansible to trigger Terraform; this is using Terraform to run other IaC tools – which some providers support.

  As this functionality is built into each provider, and each provider is a separate project away from the core Terraform development, you may see that functionality between different providers is very different.

  If you have to take this route, please consult your provider's documentation and, where necessary, examine the issues logged in its GitHub repo to see whether any problems have been reported with the functionality.

As you have seen in this section, much of the advice is similar to the general advice we covered at the start of the chapter. Let us see if this trend continues for Ansible.

Ansible – best practices and troubleshooting

At this point in the chapter, you know the drill by now: we are going start by discussing best practices, but this time putting an Ansible spin on them.

Ansible – best practices

Here are some of the best practices for approaching your Ansible playbooks:

- **Organize your playbooks with roles**: Use roles to group related tasks, variables, files, and templates, making your playbooks easier to understand and maintain.

For more information on this, see *Chapter 6, Building upon the Foundations*, where we discussed roles and Ansible Galaxy in more detail – this also leads into our next bit of advice.

- **Keep playbooks modular and reusable**: Write modular playbooks and tasks that can be reused in different scenarios to minimize duplication and improve maintainability.

This is where we start to differ slightly from Terraform, as Ansible can also be used to access both Linux or Windows hosts and execute commands on them, so reusable code for everyday tasks such as installing Apache, enabling **Internet Information Services (IIS)**, or even just patching the operating systems you are targetting will be useful.

- **Use version control**: Keep your Ansible playbook and configurations in a version control system such as Git to track changes and encourage/enable collaboration amongst your team members.

- **Employ a consistent naming convention**: Adopt a clear and consistent naming convention for tasks, files, templates, and especially variables to make it easier for other team members to pick up and follow your playbook quickly.

- **Use a dynamic inventory**: This is not something we have touched upon so far, but when Ansible targets a host's operating system, it uses an inventory file, which is a list of hosts to interact with.

Instead of hardcoding host details in an inventory file, you can use a dynamic inventory script to discover and manage resources in your environment automatically. There are scripts for most providers that typically work on tags to discover what to target.

Let's imagine your Ansible playbook launches half a dozen virtual machines in your chosen cloud. If you were to tag them with `Role:Web`, then you could use a dynamic inventory script to search the cloud provider for all virtual machines tagged with `Role` of `Web` and build up a list of IP addresses to run your playbook against.

- **Implement idempotence**: Ensure your tasks are idempotent, which means they can be executed multiple times without producing unexpected results or side effects.

If your Ansible playbook deals exclusively with just infrastructure code, then this should be straightforward, as much of this logic is handled by the APIs with which the modules will interact.

However, if you are targeting operating systems, this becomes important, as you want to avoid anything unexpected happening across potentially several hosts.

- **Secure sensitive data with Ansible Vault**: I have left this one until the end. Ansible has a built-in secrets management system called Ansible Vault, which allows you to encrypt sensitive data, such as passwords and API keys, to protect them from unauthorized access.

As well as commands such as `ansible-playbook` and `ansible-galaxy`, Ansible also ships with `ansible-vault`.

This command can encrypt and decrypt both entire files and simple strings. In the following example, we will look at encrypting a string:

```
$ ansible-vault encrypt_string --vault-id @prompt HelloWorld
```

Running the command will prompt for a new password and confirmation of the password. Once entered, it will encrypt the specified text, which is super-secret `HelloWorld`, and give you something that looks like the following:

```
New vault password (default):
Confirm new vault password (default):
Encryption successful
!vault |
$ANSIBLE_VAULT;1.1;AES256
35373665396163313561373336306261346264323638616664383766316464644
3964353266656632
63653733333837343761376563396231656635376339656330a39613333653635
3036346133393437
37383534653362306438653034383266383132393966383063666633031396439
6338326462373532
33326533564633839390a636538626261393630323733643135643339303333334
6638353039396439
3736
```

Now that you have the encrypted string, you can use it in your playbook file as in the following example (pease note that the spacing has been removed to make it easier to read):

```
---
- name: Ansible Vault Example
  hosts: localhost
  gather_facts: false
  vars:
    some_secret: !vault |
          $ANSIBLE_VAULT;1.1;AES256
6434626163656236536532663830336531633534303166643961623666333631
6361336466353461
62393331663266346363373331333039393064653731390a62656337383432
6163313133313039
3735376361353936383763623734363139336532376339323562633432356137
3434303531653831
346439393938653831370a6666656562626566643235303936333462356261336362
6532623666333565
3062
  tasks:
    - name: print the secure variable
      debug:
        var: some_secret
```

There is a copy of the preceding code in the GitHub repository that accompanies this title. To run the playbook, we need to tweak our `ansible-playbook` command slightly:

```
$ ansible-playbook --vault-id @prompt site.yml
```

Assuming you enter the correct password for the vault, this should give you something like the following output:

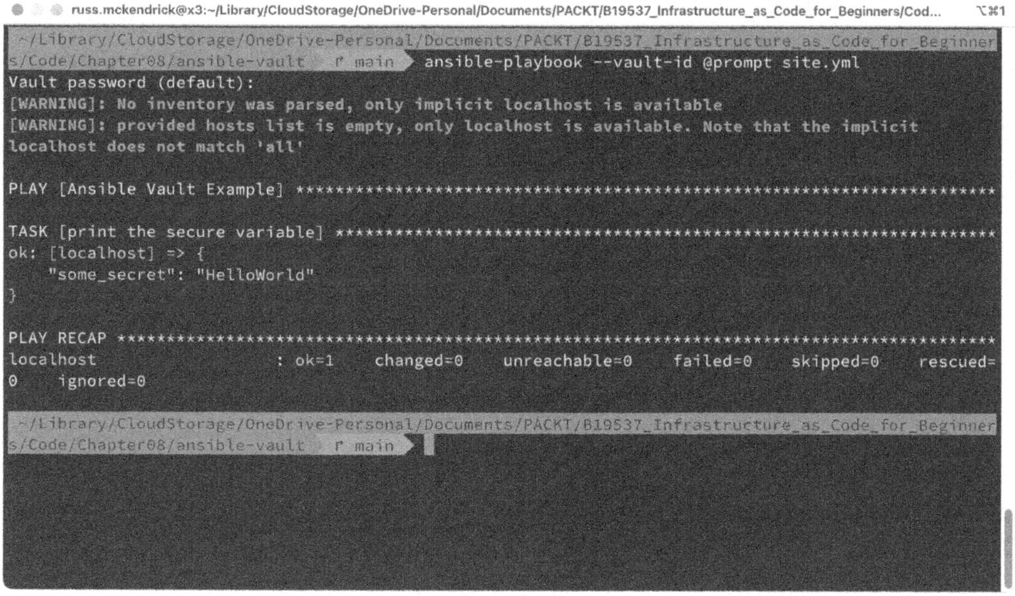

Figure 8.1 – Running the playbook and viewing the secret

If you promise not to tell anyone, the password for the playbook is `password`, so you can run the playbook in the repo yourself.

Ansible Vault can also encrypt entire files, meaning you can include files such as a private key for **Secure Sockets Layer** (**SSL**) certificates. Alternatively, we could use `base64` to encode a binary file as text and then use a vault to encrypt the encoded context, as Ansible has built-in functions for decoding `base64`.

So how is this any better than using a secret management tool? Well, it could be less complicated – you could use your secret management tool to store the password for Ansible Vault and then embed the rest of your secrets in your repo.

Now that we have covered some best practices, let's talk about troubleshooting.

Ansible – troubleshooting

What follows are some troubleshooting tips for Ansible, and a lot of the general ones also apply:

- **Use the debug module**: When writing your playbooks, use Ansible's debug module to display variables, messages, or task output, helping you identify issues in your code.

  This is extremely helpful when trying to find out the contents of a variable or the output of a task; the following example playbook uses the debug module to output the contents of the ansible_facts variable:

  ```
  ---
  - hosts: localhost
    gather_facts: yes
    tasks:
      - name: Print all the facts
        debug:
          var: ansible_facts
  ```

  Running the playbook using ansible-playbook site.yml should show you information about your host.

- **Increase verbosity levels**: Ansible hides quite a lot of information when you run ansible-playbook; you can add the -v, -vv, or -vvv options to increase the verbosity of the output, providing more insight into what's happening during execution.

- **Check your YAML syntax**: I have lost countless hours looking at a problem only to find I haven't formatted the YAML in my playbook correctly. Save yourself some time and validate your YAML files with a linter or online validator to catch any formatting or syntax errors.

- **Review the failed and skipped task summaries**: Examine the *failed* and *skipped* task summaries at the end of a playbook run to identify tasks that did not execute as expected; Ansible may not completely stop execution on a failed task, so ensure that you pay attention to your playbook runs as you may have problems and not immediately realize it.

- **Verify file and directory permissions**: Ensure that the appropriate file and directory permissions are set for your Ansible files and target hosts, allowing the required access for execution.

  For example, if you are using SSH to access a host after it has been launched, ensure that the permissions on your local machine for things such as your SSH key are correct, or your Ansible playbook run may fail.

As you can see, with the addition of managing workloads within the target hosts rather than just the infrastructure, there are a few more considerations with Ansible than with a tool such as Terraform.

Summary

We have discussed a lot in this chapter; we have talked about several similar concepts but have taken slightly different approaches depending on the tool we chose.

For me, the biggest takeaways from this chapter are as follows:

- **Version control**: Use version control to track changes and collaborate with your team and colleagues easily.

- **Documentation and consistency**: Ensure that your infrastructure code is well documented and has been written in line with your style guides or other IaC projects – no one wants to pick up messy or undocumented code during a crisis.

- **Keep an eye on the content**: Ensure you do not expose passwords, keys, or other sensitive content by checking it into your version control system. A lot of the IaC we have spoken about is designed to be human-readable, and that is the last thing you want for sensitive information.

- **Please keep it simple**: Believe me, it is very easy to go down a rabbit hole and create some very complex, and some would say overkill, IaC projects. From experience, these types of projects always end up causing more problems than they solve. They are challenging to maintain and for other team members to pick up and work with if they end up inheriting them – keep things simple and follow the previously listed takeaways.

In our next and final chapter, we are going to take a look at three other IaC tools, including two native tools from the cloud providers Microsoft Azure and Amazon Web Services, before then discussing what your next steps with IaC could be.

9

Exploring Alternative Infrastructure-as-Code Tools

Welcome to the final chapter of our **Infrastructure as Code (IaC)** journey! By now, you should be familiar with the basics of IaC and have gained hands-on experience with Terraform and Ansible.

As you progress in your career, being aware of and adept with tools in the market is crucial. This chapter aims to expand your IaC toolset by introducing you to three additional tools: **Pulumi**, **Azure Bicep**, and **AWS CloudFormation**.

While the previous tools we explored were cloud-agnostic, Azure Bicep and AWS CloudFormation are specific to their respective cloud providers. On the other hand, Pulumi sets itself apart by enabling you to use familiar programming languages such as Python to define and manage your infrastructure in actual code.

In this chapter, we are going to cover the following topics:

- Getting a hands-on understanding of Pulumi
- Getting hands-on knowledge of Azure Bicep
- Getting hands-on expertise in AWS CloudFormation

Before discussing the next steps in your IaC journey, as we have a lot to cover in this chapter, let's dive straight in and discuss Pulumi, which is, at the time of writing, the new kid on the block when it comes to IaC tools.

Technical requirements

The source code for this chapter is available here: `https://github.com/PacktPublishing/Infrastructure-as-Code-for-Beginners-/tree/main/Chapter09`.

Getting hands-on with Pulumi

So, what is Pulumi, and why has it yet to be mentioned up until now?

Pulumi is an open source IaC platform that allows developers to define, provision, and manage cloud infrastructure; however, rather than using a descriptive language with YAML (Ansible) or HCL (Terraform), it allows you to use popular programming languages such as JavaScript, TypeScript, Python, Go, and C#, as well as YAML for non-programmers.

With Pulumi, you can build, manage, and deploy IaC more familiarly and expressively, making it easier to reason about complex cloud architectures.

Pulumi supports popular cloud providers such as AWS, Azure, and Google Cloud. There is also support for tools such as Kubernetes, among others, all of which enable you to define and manage resources across multiple platforms using a single tool.

Great, you may be thinking to yourself – but why hasn't it been mentioned until now?

The answer is that it should be considered something other than a beginner's tool – given the number of different ways you can interact with it, it can be highly complex. It would require a dedicated book to do more than scratch the surface.

> **Information**
>
> Links to the instructions on installing Pulumi can be found in the *Further reading* section at the end of this chapter if you would like to follow along.

To give you an idea of how you would use Pulumi, let's look at launching a few resources in Microsoft Azure, as we did in the early examples of Terraform and Ansible we covered; we will be creating a resource group and a storage account.

We will start by using YAML and then look at the same deployment in Python.

Using Pulumi and YAML

We have two files, both of which can be found in the GitHub repository accompanying this book. The first file, which defines some environment-specific configurations, is called `Pulumi.dev.yaml` and, for our example, contains the following code:

```
config:
  azure-native:location: UKSouth
```

As you can see, all we are doing is defining the default `location` to be used by the Azure Native provider.

The second of the two files is called `Pulumi.yaml`, and it starts by defining some information and settings for our project:

```
name: pulumi-yaml
runtime: yaml
description: A minimal Azure Native Pulumi YAML program
outputs:
  primaryStorageKey: ${storageAccountKeys.keys[0].value}
```

The first three lines, `name`, `runtime`, and `description`, all define some basic meta information about our deployment.

The following two lines define the output, which in our case will be the primary key of the storage account that will be created.

Here, we are defining an output variable of `primaryStorageKey`, which is taking its value from a variable we will define at the end; this variable will contain the outpoint of a function we will be running once the storage account has been created.

Now that we have the basics in place, let's define the resources using a resource block, starting with the Azure Resource Group:

```
resources:
  resourceGroup:
    type: azure-native:resources:ResourceGroup
    properties:
      resourceGroupName: rg-pulumi-yaml
```

As you can see, this is not too dissimilar structurally from Terraform and Ansible – here, we are defining a resource that will be referred to as `resourceGroup`, which has a type of `azure-native:resources:ResourceGroup`, before finally setting a single property that contains the `resourceGroupName` key.

Now that the Resource Group has been defined, we can add the storage account resource, which we are going to refer to as `sa`:

```
sa:
  type: azure-native:storage:StorageAccount
  properties:
    kind: StorageV2
    resourceGroupName: ${resourceGroup.name}
    sku:
      name: Standard_LRS
```

Again, it follows the same pattern as before; we set the resource reference and the type of resource we want to create and then define our `properties`.

In this case, rather than the name of the resource, which Pulumi will create for us, we are passing in a `sku` name, the kind of storage account we want to create (via `kind`), and the `resourceGroupName` key to add the resource.

To do this, we must use the `${resourceGroup.name}` variable, which takes the name of the resource group we referenced as `resourceGroup`. Like in Terraform, this ensures that the resource group is created before the storage account.

The final part of the `Pulumi.yaml` file sets the `storageAccountKeys` variable, which is used by the output section we start at the start of the file.

To do this, we need to define a `variables` section:

```
variables:
  storageAccountKeys:
    fn::azure-native:storage:listStorageAccountKeys:
      accountName: ${sa.name}
      resourceGroupName: ${resourceGroup.name}
```

Here, we are setting the function (`fn`), which is an `azure-native` one that deals with `storage` and is called `listStorageAccountKeys`. It requires two inputs – `accountName`, which we pass in using `${sa.name}` and, as most things need in Azure, the `resourceGroupName` key. So, as before, we pass this in programmatically by using the `${resourceGroup.name}` variable.

Now that we have all of the code, let's launch the resources. To do this, we need to issue the following command:

```
$ pulumi up -c Pulumi.dev.yaml
```

This is where things get a little different than Terraform and Ansible; the first thing that happens is that you are asked to log in, as shown in the following screenshot:

Figure 9.1 – Running Pulumi up for the first time

Follow the onscreen prompts and press the *Enter* key to be taken to the login page. Here, you can sign up or log in using one of the many supported identity providers; I used GitHub. Once you have logged in or signed up, you should get the option to create a stack:

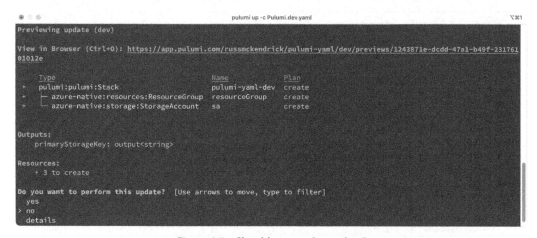

Figure 9.2 – Time to create a stack

Once you have created your stack, Pulumi will run a check against your code and give you the option to deploy the update. In this case, this is going to create three resources – the two in Azure and our output:

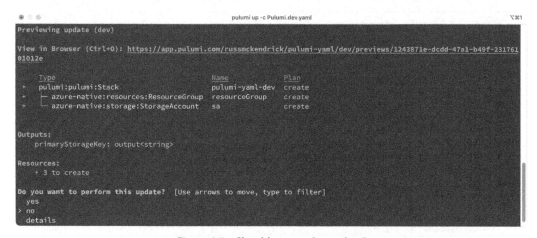

Figure 9.3 – Should we run the update?

If you use the arrow keys to select **yes** and then hit the *Enter* key, Pulumi will deploy the resources:

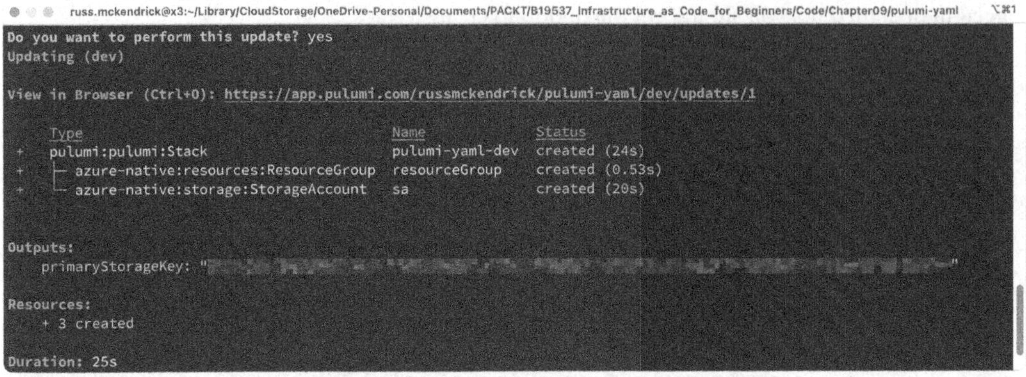

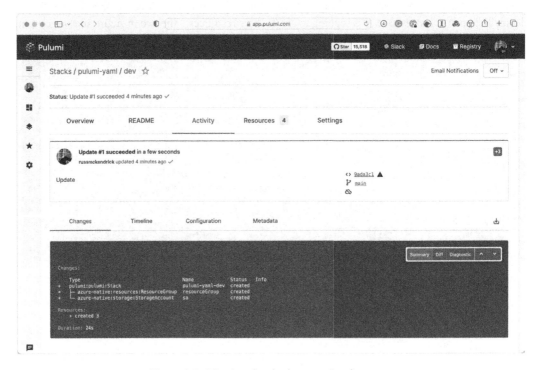

Figure 9.4 – The deployment has been completed

As shown in the preceding output, we have the output (which I have blurred out the value of) and an overview of the deployment. The eagle-eyed among you may have also noticed a URL – clicking on it opens an overview of the deployment in your browser. For me, this looked like this:

Figure 9.5 – Viewing the deployment in a browser

I recommend having a look around your stack in the browser. Once you have finished, you can remove the resources by running the following command:

```
$ pulumi destroy
```

This will remove the Azure resources but not the stack on the Pulumi website.

Now, let's look at deploying the same resources again, but instead of using YAML, we will use Python.

Using Pulumi and Python

This, as you may have already guessed, is where things start to get a little more advanced.

In the repository that accompanies the book, you will find several files; these are as follows:

- .gitignore: This contains entries for the venv and __pycache__ folders, which we do not need to check into version control

- __main__.py: This is the main Python code; we will cover this in more detail shortly

- Pulumi.dev.yaml: This contains the environment config and has the same contents as we used YAML rather than Python

- Pulumi.yaml: This contains the basic metadata for our deployment

- requirements.txt: Like most Python scripts, there are external dependencies; this file lists these so that they can be installed using pip

Let's start by looking at the requirements.txt file. As mentioned, this contains the dependencies needed to run our Python code:

```
pulumi>=3.0.0,<4.0.0
pulumi-azure-native>=1.0.0,<2.0.0
```

As you can see, there are just two dependencies – Pulumi and the Azure Native provider.

As already mentioned, we have the Pulumi.yaml file. Even though we are using Python, it contains the basic information and settings for the project:

```
name: pulumi-python
runtime:
  name: python
  options:
    virtualenv: venv
description: A minimal Azure Native Python Pulumi program
```

As you can see, `runtime` is now `python`, and some settings define the folder where the Python virtual environment (`virtualenv`) will be stored. In our case, this is `venv` and is in the same folder as the rest of our project files.

The final file is the `__main__.py` file and is where our resources are defined. The first part of the file imports the Python libraries needed to deploy the resources:

```
"""An Azure RM Python Pulumi program"""
import pulumi
from pulumi_azure_native import storage
from pulumi_azure_native import resources
```

As you can see, of the two dependencies defined in the `requirements.txt` file, we are importing all of the `pulumi` library; however, `storage` and `resources` from the `pulumi_azure_native` library as the Resource Group and storage account, respectively, are the only two resources that we are launching. Therefore, we do not need to load the entire library.

Next up, we must define the Resource Group:

```
resource_group = resources.ResourceGroup(
    "resource_group",
    resource_group_name="rg-pulumi-python",
)
```

I wouldn't call myself a Python programmer – I know enough to be dangerous and run the basics – but I am sure you will agree that the code looks simple enough.

Now, let's define the storage account:

```
account = storage.StorageAccount(
    "sa",
    resource_group_name=resource_group.name,
    sku=storage.SkuArgs(
        name=storage.SkuName.STANDARD_LRS,
    ),
    kind=storage.Kind.STORAGE_V2,
)
```

Again, it is a little different from what we dealt with when defining our infrastructure in YAML or HCL.

But again, it is simple enough to follow what is going on within the code, mainly as we have already deployed this same project using Pulumi and YAML.

This also means you should have an idea of what is coming up next – that is, the function to grab the storage account key:

```
primary_key = (
    pulumi.Output.all(resource_group.name, account.name)
    .apply(
        lambda args: storage.list_storage_account_keys(
            resource_group_name=args[0], account_name=args[1]
        )
    )
    .apply(lambda accountKeys: accountKeys.keys[0].value)
)
pulumi.export("primary_storage_key", primary_key)
```

This is where things turn a little more into a traditional Python script; it is still relatively straightforward to follow what is going on but if, like me, you are not a Python developer, you may find it a little more challenging to write the preceding code from scratch.

Let's try deploying the code. To do this, we simply use the same command as before:

```
$ pulumi up -c Pulumi.dev.yaml
```

You will notice some differences when you first run the command:

Figure 9.6 – Installing the dependencies

As you may have already guessed, first, the dependencies defined in the `requirements.txt` file must be installed.

Once our dependencies have been installed, we drop back to the same options as we were presented with when we deployed the YAML version of the project:

```
● ○ ○                          pulumi up -c Pulumi.dev.yaml                              ⌥⌘1
lumi-python  ⎇ main   pulumi up -c Pulumi.dev.yaml
Previewing update (dev)

View in Browser (Ctrl+O): https://app.pulumi.com/russmckendrick/pulumi-python/dev/previews/3a31e509-4fbc-49d8-8479-c5bd
b031df19

    Type                                   Name                  Plan
 +  pulumi:pulumi:Stack                    pulumi-python-dev     create
 +   └─ azure-native:resources:ResourceGroup   resource_group    create
 +   └─ azure-native:storage:StorageAccount    sa                create


Outputs:
    primary_storage_key: output<string>

Resources:
    + 3 to create

Do you want to perform this update?  [Use arrows to move, type to filter]
  yes
> no
  details
```

Figure 9.7 – Back in familiar territory

Again, you get a URL to view your stack on the Pulumi website, and you can terminate the resources by running the `pulumi destroy` command.

So, why do this?

Most of the audience for this book, I imagine, comes from an operations or system administration background rather than a programming one – this means you are more familiar with working with configuration files of all types and understand the steps you need to take to deploy your infrastructure.

Pulumi aims to appeal to people from that background as well as developers by offering them a way of defining their infrastructure in a language that is familiar to them; as you may recall from the start of this section, JavaScript, TypeScript, Python, Go, and C# are all supported.

Another advantage is that you can move your IaC into your existing build and deployment pipelines. For example, let's say you have a mature C# build, test, and deployment workflow. If you are using Pulumi, you should be able to introduce your IaC into the process quickly.

As mentioned at the start of this section, we have yet to begin to unlock the power of Pulumi in this section – but I am sure you will agree that it opens up many possibilities when approaching your IaC deployments.

Now that we have looked at the last of the cloud-agnostic tools, let's look at the two cloud-native tools before we finish.

Getting hands-on knowledge of Azure Bicep

Azure Bicep is the first of the two cloud-specific IaC tools we will be looking at in this chapter. For quite a while, if you wanted to use the native tool provided by Microsoft, you would need to write an ARM template.

When we discussed Microsoft Azure in *Chapter 4, Deploying to Microsoft Azure*, we stated that ARM is short for Azure Resource Manager – that is, the API that powers all of Azure. You will have been using ARM when using the Azure portal, command-line tools, PowerShell, or any IaC tool we have covered to launch or manage your Microsoft Azure resources.

The best way I can think to describe ARM templates is that they are the JSON payloads that are sent to the API – I won't include an example of what an ARM template looks like as there is a lot of it, but I have included an example file called `arm-template-example.json` in the same folder as the Bicep file in the accompanying repository. As you can see, there is a lot of it; the file is just short of 120 lines of code – and all that does is define a storage account.

So, now that we have provided a quick explanation of ARM templates, let's look at Bicep.

Bicep is a **domain-specific language** (**DSL**) that employs a declarative syntax for deploying Azure resources. From that description, you may think that it sounds not that different from Terraform, and you would be right – let's dive straight into the Bicep code, which is in a file called `main.bicep`.

Working through the Bicep file

The first part of our Bicep code sets the parameters, of which we are going to set three up, starting with the type of storage account we will be launching:

```
@description('Storage Account type')
@allowed([
    'Premium_LRS'
    'Premium_ZRS'
    'Standard_GRS'
    'Standard_GZRS'
    'Standard_LRS'
    'Standard_RAGRS'
    'Standard_RAGZRS'
    'Standard_ZRS'
])
param storageAccountType string = 'Standard_LRS'
```

As you can see, here, we provide an array of the `allowed` possible values before defining a parameter (`param`) called `storageAccountType` with a string value of `Standard_LRS`. This means that if we override the default parameter at runtime, it will only accept one of the allowed parameters rather than just any old string.

The second and third parameters are as follows:

```
@description('The storage account location.')
param location string = resourceGroup().location
```

The former sets the `location` parameter by inheriting the location of the Resource Group; we also use the ID of the Resource Group to generate a unique string for the storage account's name:

```
@description('The name of the storage account')
param storageAccountName string = 'sa${uniqueString(resourceGroup().
id)}'
```

The `uniqueString` function takes the Resource Group ID, which in itself is unique to your deployment, and uses that to generate a string. This means that you know a consistent string is being generated rather than a random string each time the Bicep code is executed. However, as it is based on the Resource Group ID that is unique to your deployment, you should never have two strings that are the same.

So, now that we have defined our three parameters, we can add the code to create the storage account resource.

The block to do this looks like this:

```
resource sa 'Microsoft.Storage/storageAccounts@2022-09-01' = {
  name: storageAccountName
  location: location
  sku: {
    name: storageAccountType
  }
  kind: 'StorageV2'
  properties: {}
}
```

Here, we are creating a `resource` block that is going to be referred to as sa; it uses the `Microsoft.Storage/storageAccounts@2022-09-01` API endpoint. We are also passing in the parameters for name, `location`, and `sku`.

The final two lines of code set some output, which is the storage account name and the ID:

```
output storageAccountName string = storageAccountName
output storageAccountId string = sa.id
```

You may have noticed that's something missing… have you guessed what it is?

Deploying the Bicep file

If you guessed the following, then you would be correct:

"Hang on a minute; we are referencing a Resource Group, but we are not defining a block for one."

By default, both Bicep and ARM templates expect you to deploy into a resource rather than have one defined within the Bicep file.

Another thing that you may have noticed is that I haven't given any instructions on how to install Azure Bicep.

The reason for this is that Bicep is built into the Azure CLI, which we will also use to create a resource group. Do this by running the following command:

```
$ az group create -l uksouth -n rg-bicep-example
```

When I ran the command, I got the following output:

```
{
  "id": "/subscriptions/3e3c9f50-1a27-4e7e-af2e-e0d3f3e4a8f4/
resourceGroups/rg-bicep-example",
  "location": "uksouth",
  "managedBy": null,
  "name": "rg-bicep-example",
  "properties": {
    "provisioningState": "Succeeded"
  },
  "tags": null,
  "type": "Microsoft.Resources/resourceGroups"
}
```

The command will create a resource group called `rg-bicep-example` in the UK South region, which we can now deploy our Bicep file into by running the following code:

```
$ az deployment group create --resource-group rg-bicep-example
--template-file main.bicep
```

This will output quite a bit of information, but the two import bits we are interested in are the outputs. For me, these looked like this:

```
"outputs": {
  "storageAccountId": {
    "type": "String",
    "value": "/subscriptions/ce7aa0b9-3545-4104-99dc-d4d082339a05/
resourceGroups/rg-bicep-example/providers/Microsoft.Storage/
storageAccounts/saljkmvlrqknl2y"
  },
  "storageAccountName": {
```

```
    "type": "String",
    "value": "saljkmvlrqknl2y"
  }
},
```

As you can see, `storageAccountId` and `storageAccountName` are visible.

> **Important**
>
> The following command deletes the whole resource group and everything within it, so please be careful and only run if you want everything to be deleted.

You can remove the resources we launched with Bicep by running the following command:

```
$ az group delete -n rg-bicep-example
```

Again, this section wasn't planned to be a deep dive into Bicep; I wanted to give a basic example to show you how IaC tools are not just limited to the "big two" of Terraform and Ansible. We have yet to come close to scratching the surface of what is possible with Bicep.

Since Microsoft first launched the alpha release of Bicep in August 2020, it has quickly grown and become a first-class citizen within the Azure ecosystem; for example, all of the official Azure documentation now includes references to and examples of both ARM templates and Bicep code for launching and interacting with your Azure resources.

Also, as we have experienced, it is built directly into the Azure CLI, meaning you already have it at your disposal if you are already working with Microsoft Azure.

Before we discuss why you should use Bicep over one of the other tools, let's look at the other cloud-native option – AWS CloudFormation.

Getting hands-on with AWS CloudFormation

AWS CloudFormation is a service provided by Amazon Web Services that enables you to manage and provision your AWS resources using templates.

Of all the tools we have looked at in this book, AWS CloudFormation is the oldest, with its original public release in May 2010. Also, in the description, I described it as a service that uses templates – this all makes the approach slightly different than the other tools we have covered.

CloudFormation uses JSON or YAML templates to describe your desired AWS resources and their configurations. These templates define a stack, which is a collection of related resources that can be created, updated, or deleted together.

It provides automatic rollback and drift detection capabilities to help you maintain the desired state of your infrastructure. CloudFormation can automatically revert to the previous working state if a stack update fails. Drift detection allows you to identify and correct discrepancies between the actual infrastructure and the desired state defined in the template.

Also, before deploying a stack, you can estimate the cost of the resources defined in your template. Additionally, you can use tags to categorize and track costs associated with specific resources, projects, or environments.

We will look at deploying a single Amazon S3 bucket using both the AWS command line and the AWS Management Console.

AWS CloudFormation template

First, let's look at the template file we will be using. I prefer to use YAML over JSON as it is much easier to read and follow what is going on.

The template we will be using is split into four small sections. The template is a small 20-line file; I have seen templates containing several hundred lines of code, so this is the most basic example we can use.

The start of the template contains some basic information, including a description of what the template does and which format to use:

```
AWSTemplateFormatVersion: "2010-09-09"
Description: Creates a basic S3 bucket using CloudFormation
```

Next, we must set some `Parameters`; in our case, this is just going to be `BucketName`:

```
Parameters:
  BucketName: { Type: String, Default: "my-example-bucket-name" }
```

Next, we have the resources, where we define our S3 bucket:

```
Resources:
  ExampleBucket:
    Type: "AWS::S3::Bucket"
    Properties:
      BucketName: !Ref BucketName
      BucketEncryption:
        ServerSideEncryptionConfiguration:
          - ServerSideEncryptionByDefault:
              SSEAlgorithm: AES256
```

Finally, we must set an output that returns the name of the bucket we created:

```
Outputs:
  ExampleBucketName:
    Description: Name of the example bucket
    Value: !Ref ExampleBucket
```

As you can see, there are some differences in how parameters are referenced; I am not a fan of using syntax such as !Ref BucketName only because the other tools, all of which came after, use a similar way of referencing parameters/variables.

Now that we have our template, let's look at using the AWS CLI to deploy the stack.

Using the AWS CLI to deploy

The AWS CLI makes it easy to deploy our template. To deploy the S3 bucket, run the following command, making sure you update the name of the bucket at the end of the command to your own. This is because bucket names need to be unique:

```
$ aws cloudformation create-stack --stack-name iaccloudform
--template-body file://cftemplate.yaml --parameters
ParameterKey=BucketName,ParameterValue=iac230404
```

Once deployed, you should get some output that looks like this:

```
{
    "StackId": "arn:aws:cloudformation:us-west-2:687011238589:stack/
iaccloudform/ca605040-d2fa-11ed-84fd-027270021b81"
}
```

We have just deployed the stack but not the resources – the stack, which is an AWS service, will be deploying those for you in the background.

To delete the stack we just launched and the resources managed by it, run the following command:

```
$ aws cloudformation delete-stack --stack-name iaccloudform
```

Let's see what creating a stack in the AWS Management Console looks like.

Using the AWS Management Console to deploy

Once logged into the AWS Management Console, go to CloudFormation and click the **Create stack** button.

The first step of creating a stack is to define your template. Since you already have one, ensure that **Template is ready** is selected. Then, select the **Upload a template file** option and press the **Choose file** button to upload your file:

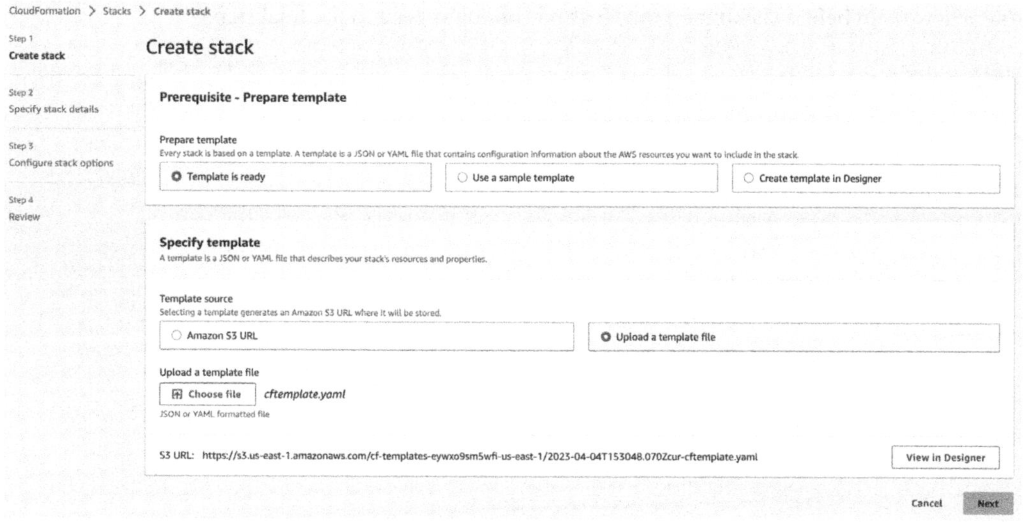

Figure 9.8 – Completing step one

The second step is to provide some details about the stack and update any parameters:

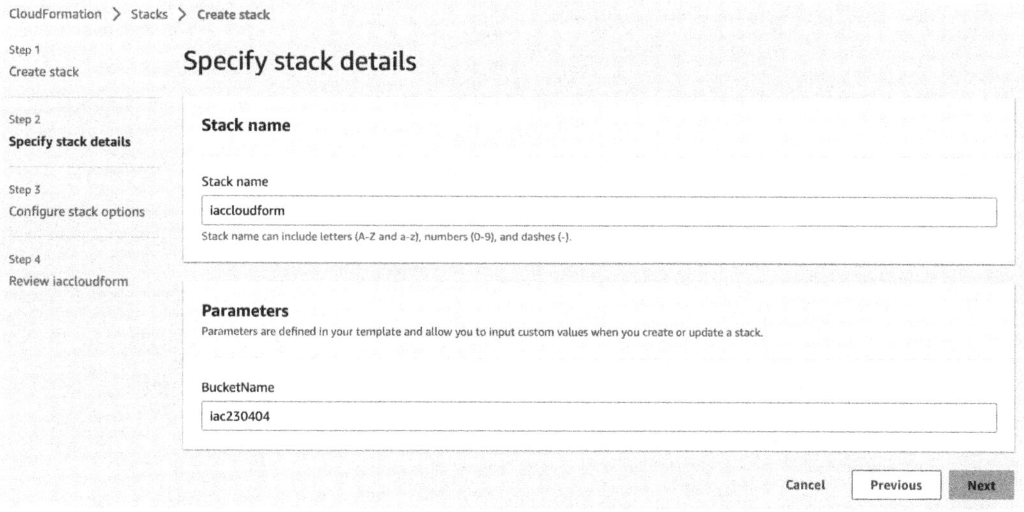

Figure 9.9 – Entering details of the stack

Step three is where you configure options for the stack; here, you can define tags and permissions and control the actions taken should a deployment fail.

For our deployment, you can leave all of the options at their defaults – however, I recommend that you review them before clicking on the **Next** button to proceed to the final step:

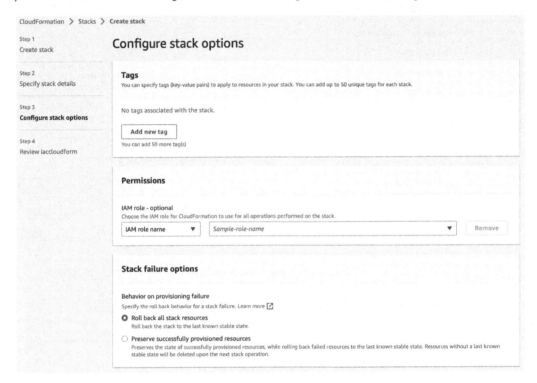

Figure 9.10 – Reviewing the stack options

The final step is to review your stack before clicking the **Submit** button, triggering the stack's creation:

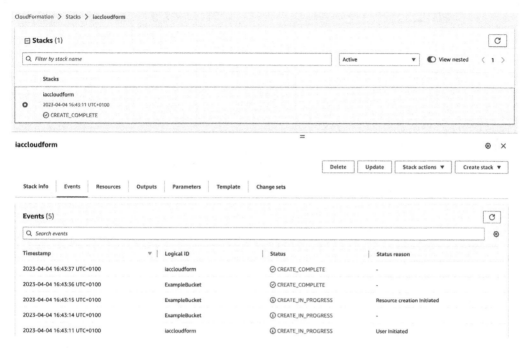

Figure 9.11 – Reviewing the deployment

From here, you can review your resources. Once you have finished, clicking on **Delete** will remove the stack.

One thing that you will have noticed is that there are some sample templates, as well as a template designer. Loading one of the samples into the designer gives you a graphical view of the template and a drag-and-drop interface that you can use to design your templates:

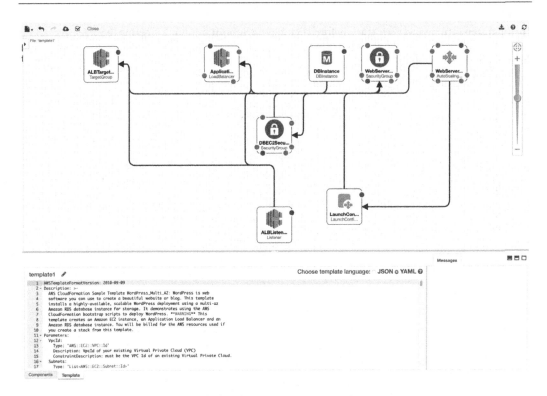

Figure 9.12 – Designing your template

As you can see, there are options to export your finished template as JSON or YAML; in our example, there are just over 700 lines of code in the YAML file.

This is the biggest reason you want to use the designer and AWS Management Console.

AWS CloudFormation can quickly become very complex, and it doesn't lend itself well to sitting in front of an empty file and starting to code – I find it very overwhelming.

Summary

In this final chapter, we looked at three additional IaC tools, all of which work slightly differently from the two primary tools we looked at in previous chapters. So, why would you choose these over Terraform or Ansible?

In *Chapter 2, Ansible and Terraform beyond the Documentation*, we concluded that you should choose the best tool for the job rather than trying to fit your project to the tool; the same goes for the tools we have discussed in this chapter.

When planning your IaC project, having an excellent working knowledge of more than one tool is always a bonus; throughout this book, there have been occasions when either Terraform or Ansible hasn't supported a task we were trying to perform, so we had to use the built-in tools that provide support for the target cloud's API.

If you have a project in Azure, for example, where coverage for the latest services in the tools may be behind by several months, then using Azure Bicep may be the best choice as you know you are exclusively targeting Azure; Bicep has day 1 support for 99.9% of all new Azure services.

Likewise, you may have to work alongside developers who want to bring your deployments into their existing processes and procedures; therefore, using Pulumi may be more suitable than introducing one of the other tools.

So, what should your immediate next steps be?

Suppose you have access to a lab or a free cloud account. In that case, I recommend choosing a typical deployment and working through the steps covered in the first three chapters to define your project and then execute your IaC project.

Before starting, ensure you know what your end deployment will look like and how it needs to be configured. From there, you should be able to break it down into tasks, which will give you an idea of the dependencies.

Once you have an idea of the tasks and dependencies, this should allow you to work out the order in which the tasks would need to be executed – this is where you should choose which tool to use. However, you shouldn't do this before as you need to know if you need a decorative or imperative tool and you must know the compatibility and service support each tool would need for your deployment to succeed.

Once you know what it is you are deploying, in which order, and using which tool, you can open a blank file and start to write your code.

I recommend writing some code and doing a test deployment – to deal with any issues – and then terminating the resources once you have resolved the problem.

Do not leave it to the end to try and debug your code. Also, ensure that when you do test deployments, you remove them – otherwise, you may end up introducing dependency issues into your deployments as resources may already exist, and therefore any issues or errors within your code may not reveal themselves.

Expect a lot of trial and error, especially if you are new to IaC deployments. Many considerations may not be completely apparent if you are used to deploying resources using the Azure portal or AWS Management Console, as these interfaces take a lot of the heavy lifting from you and do quite a bit of work in the background to make the process of launching your resources as smooth as possible.

Finally, once you have something up and running, make sure you show as many people as possible, give them access to your code if appropriate – show them it being deployed, try and sell them on the benefits of taking an IaC approach to their projects, and be as supportive as possible.

Thank you for allowing me to accompany you on this journey; I wish you every success with your projects.

Further reading

Pulumi:

- Download and installation instructions: `https://www.pulumi.com/docs/get-started/install/`

- Getting started with Azure: `https://www.pulumi.com/docs/get-started/azure/`

- Getting started with AWS: `https://www.pulumi.com/docs/get-started/aws/`

- Importing your infrastructure and converting your existing IaC: `https://www.pulumi.com/docs/guides/adopting/`

Azure Bicep:

- Bicep overview: `https://learn.microsoft.com/en-us/azure/azure-resource-manager/bicep/overview?tabs=bicep`

- Download and install: `https://learn.microsoft.com/en-us/azure/azure-resource-manager/bicep/install`

- Learn Bicep Live: `https://learn.microsoft.com/en-us/events/learn-events/learnlive-iac-and-bicep/`

AWS CloudFormation:

- Product page: `https://aws.amazon.com/cloudformation/`

- Full documentation: `https://docs.aws.amazon.com/AmazonCloudFront/latest/DeveloperGuide/Introduction.html`

Index

A

admin host 39
admin virtual machine 66-69
Admin Virtual Machine role 80
Amazon EC2 86, 89
Amazon Machine Image (AMI) 97
Amazon Web Services (AWS) 85-87
Ansible 29, 30, 126
  best practices 168-171
  code, reviewing 73
  code, writing 89-91
  example 30-33
  infrastructure, deploying 73, 89-91
  running, with GitHub Actions 155-157
  troubleshooting tips 172
  versus Terraform 32
Ansible deployments
  versus Terraform deployments 126-129
Ansible playbook
  Auto Scaling group 104
  EC2 instance 104
  EFS instance 104
  Elastic load balancer 104
  RDS instance 104
  running 81, 102, 103

  security group 105
  Virtual Private Cloud 106
Ansible playbook roles 91
  Admin Virtual Machine role 80
  AWS ASG role 100, 101
  AWS Database role 96
  AWS Network role 92
  AWS Network role, creating 92-95
  AWS Storage role 95
  AWS Storage role, creating 96
  AWS VM Admin role 96
  AWS VM Admin role, creating 97-100
  MySQL role 78, 79
  output role 80, 102
  overview 74-76
  random role, creating 92
  resource group role 76
  Storage roles 77, 78
  Virtual Network role 76, 77
  Web Virtual Machine Scale Set role 80
Ansible Vault 169
application programming
  interfaces (APIs) 86
automation solutions 16
Auto Scaling groups (ASGs) 89, 104
AWS ASG role 100, 101
AWS CLI installation 87

AWS CloudFormation 188
  AWS CLI, used for deploying stack 190
  AWS Management Console, used
    for deploying stack 190-194
  template 189, 190
  working with 188, 189
AWS Database role 96
AWS Management Console
  reference link 107
AWS Network role 92-95
AWS Storage role 95, 96
AWS VM Admin role 96-100
Azure Application Gateway 51
Azure Bicep 185
  code 185
  file, deploying 187, 188
  file, working through 185, 186
Azure Database for MySQL -
  Flexible Server 52, 66
Azure Load Balancer 51, 56
Azure Private DNS 52
Azure Resource Manager (ARM) 168
Azure Storage Account/Azure Files 52
Azure Virtual Network 56-63

B

Base64 67
bootstrapping 40

C

Cattle resources 13
  versus Pets resources 13
ChatGPT 130
Chef tool 5
Claranet 132
Classless Inter-Domain Routing (CIDR) 167

Cloud Adoption Framework 54
cloud agnostic tools 117
cloud environment
  preparing, for deployment 50, 87, 88
cloud-init 42, 43
code
  reusable 131-134
collaboration 15
Command-Line Interface (CLI) 50, 87, 142
content management system (CMS) 38
Continuous Integration and Continuous
  Deployment (CI/CD) 126, 163

D

Database-as-a-Service (DBaaS) 52
database service 65, 66
declarative approach 8-11
domain-specific language (DSL) 185
dynamic block 60

E

EC2 instance 104
EFS instance 104
Elastic File System (EFS) 89
Elastic Load Balancing (ELB) 89, 104
eventual consistency 11

G

GitHub Actions 140, 142-155
  Ansible, running with 155-157
  events 140
  jobs 140
  steps 140
  Terraform, running with 141
  workflows 140

H

Hashicorp Configuration
Language (HCL) 22
high-level architecture 43, 44

I

imperative approach 8, 11-13
Infrastructure-as-a-Service (IaaS) 21, 50
Infrastructure as Code (IaC) 37, 118
advantages 15, 16
best practices 162-164
challenges 3-5
deployment 15, 16
documenting 5, 6
reasons, for incorporating to deployments 7
resources, deploying 6
scenarios 7, 8
servers, working with 4
troubleshooting tips 164, 165
Infrastructure as Code project 8
components 9
Infrastructure as Code, servers
bare-metal 4
deployment 5
virtualization 5
virtual machine configuration 5
infrastructure deployment, approaching
deployment considerations 39, 40
deployment tasks, performing 40, 41
Internet Information Services (IIS) 127, 169

L

Linux virtual machine 52
load balancing 40

low-level design
producing 51, 88, 89

M

Microsoft Azure
background knowledge 50
Microsoft Azure versus Amazon
Web Services deployments
database 121
general 119
implementing 122-125
network 119
storage 120
virtual machine (admin) 121
virtual machines with scaling (web) 122
Microsoft .NET Services 50
Microsoft SharePoint and Dynamics 50
Microsoft SQL Data Services 50
MySQL role 78, 79

N

network interface 9
Network Security Group 9, 52, 56

O

output role 80, 102

P

Pets versus Cattle resources 13
conclusion 14, 15
Platform as a Service (PaaS) 39, 50, 120
private endpoints 52
private networking 40
programming paradigms 8

Project Red Dog 50

public IP address 9, 52

Pulumi 176

   using, with Python 181-184

   using, with YAML 176-181

   working with 176

Puppet tool 5

Python

   using, with Pulumi 181-184

R

randoms role 74

RDS instance 104

Relational Database Service (RDS) 89

Resource Group 9

   creating 54-56

   naming 55

resource group role 76

resource tagging 56

S

Secure by Design (SBD) 164

Secure Shell (SSH) 21, 127, 166

Secure Sockets Layer (SSL) 171

security

   best practices 158

security group 105

shared storage 40

Simple Queue Service (SQS) 86

Simple Storage Service (S3) 86, 141, 166

Software as a Service (SaaS) 50

storage account 64, 65

Storage roles 77, 78

Subnet 9

T

Team Foundation Server (TFS) 162

Terraform 22, 126

   best practices 165-167

   code, reviewing 107

   code, writing 52, 53

   environment, deploying 113

   GitHub Actions, used for running 141

   HCL example 23

   infrastructure, deploying 52, 53, 107

   resource groups, creating 23-25

   resources error, fixing 27-29

   state files 141, 142

   storage account, adding 25-27

   troubleshooting 167, 168

   versus Ansible 32

Terraform deployments

   versus Ansible deployments 126-129

Terraform environment

   deploying 70-73

   outputs 69

   setting up 53, 54

Terraform files

   Auto Scaling group 112

   database 109

   networking 108

   output 112

   setup 107

   storage 109

   virtual machine 110, 111

   walk-through 107

Terraform Registry 54

tool

   selecting 19

tool selection, consideration

   deployment types 20

   deployment types, advantage 20

deployment types, challenges 20
deployment types, use case 20
ease of use 22
external interactions and secrets 21
infrastructure and configuration 21

U

Universally Unique Identifiers (UUIDs) 142

V

variables
  using 129-131
virtual machine 9, 52
virtual machine instances 39
Virtual Machine Scale Set 52
Virtual Network 9, 52
Virtual Network role 76, 77
Virtual Private Cloud (VPC) 89, 106
Visual Studio Code 34, 35
  extensions 35

W

Web Virtual Machine Scale Set 69
Web Virtual Machine Scale Set role 80
Windows Azure 50
Windows Remote Management (WinRM) 21
WordPress 38
  features 38
workload
  deployment, planning 37, 38

Y

YAML Ain't Markup Language 146
YAML Lint
  URL 30
Yet Another Markup Language (YAML) 146
  using, with Pulumi 176-181

Packtpub.com

Subscribe to our online digital library for full access to over 7,000 books and videos, as well as industry leading tools to help you plan your personal development and advance your career. For more information, please visit our website.

Why subscribe?

- Spend less time learning and more time coding with practical eBooks and Videos from over 4,000 industry professionals
- Improve your learning with Skill Plans built especially for you
- Get a free eBook or video every month
- Fully searchable for easy access to vital information
- Copy and paste, print, and bookmark content

Did you know that Packt offers eBook versions of every book published, with PDF and ePub files available? You can upgrade to the eBook version at packtpub.com and as a print book customer, you are entitled to a discount on the eBook copy. Get in touch with us at customercare@packtpub.com for more details.

At www.packtpub.com, you can also read a collection of free technical articles, sign up for a range of free newsletters, and receive exclusive discounts and offers on Packt books and eBooks.

Other Books You May Enjoy

If you enjoyed this book, you may be interested in these other books by Packt:

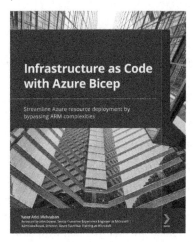

Infrastructure as Code with Azure Bicep

Yaser Adel Mehraban

ISBN: 9781801813747

- Get started with Azure Bicep and install the necessary tools
- Understand the details of how to define resources with Bicep
- Use modules to create templates for different teams in your company
- Optimize templates using expressions, conditions, and loops
- Make customizable templates using parameters, variables, and functions
- Deploy templates locally or from Azure DevOps or GitHub
- Stay on top of your IaC with best practices and industry standards

Ansible for Real-Life Automation

Gineesh Madapparambath

ISBN: 9781803235417

- Explore real-life IT automation use cases and employ Ansible for automation
- Develop playbooks with best practices for production environments
- Approach different automation use cases with the most suitable methods
- Use Ansible for infrastructure management and automate VMWare, AWS, and GCP
- Integrate Ansible with Terraform, Jenkins, OpenShift, and Kubernetes
- Manage container platforms such as Kubernetes and OpenShift with Ansible
- Get to know the Red Hat Ansible Automation Platform and its capabilities

Packt is searching for authors like you

If you're interested in becoming an author for Packt, please visit `authors.packtpub.com` and apply today. We have worked with thousands of developers and tech professionals, just like you, to help them share their insight with the global tech community. You can make a general application, apply for a specific hot topic that we are recruiting an author for, or submit your own idea.

Share Your Thoughts

Now you've finished *Infrastructure as Code for Beginners*, we'd love to hear your thoughts! Scan the QR code below to go straight to the Amazon review page for this book and share your feedback or leave a review on the site that you purchased it from.

`https://packt.link/r/1837631638`

Your review is important to us and the tech community and will help us make sure we're delivering excellent quality content.

Download a free PDF copy of this book

Thanks for purchasing this book!

Do you like to read on the go but are unable to carry your print books everywhere? Is your eBook purchase not compatible with the device of your choice?

Don't worry, now with every Packt book you get a DRM-free PDF version of that book at no cost.

Read anywhere, any place, on any device. Search, copy, and paste code from your favorite technical books directly into your application.

The perks don't stop there, you can get exclusive access to discounts, newsletters, and great free content in your inbox daily

Follow these simple steps to get the benefits:

1. Scan the QR code or visit the link below

https://packt.link/free-ebook/978-1-83763-163-6

2. Submit your proof of purchase
3. That's it! We'll send your free PDF and other benefits to your email directly

Made in the USA
Coppell, TX
20 September 2023

21790666R00122